Lecture Notes in Mechanical Engineering

Lecture Notes in Mechanical Engineering (LNME) publishes the latest developments in Mechanical Engineering—quickly, informally and with high quality. Original research reported in proceedings and post-proceedings represents the core of LNME. Volumes published in LNME embrace all aspects, subfields and new challenges of mechanical engineering.

To submit a proposal or request further information, please contact the Springer Editor of your location:

Europe, USA, Africa: Leontina Di Cecco at Leontina.dicecco@springer.com
China: Ella Zhang at ella.zhang@springer.com
India: Priya Vyas at priya.vyas@springer.com
Rest of Asia, Australia, New Zealand: Swati Meherishi
at swati.meherishi@springer.com

Topics in the series include:

- Engineering Design
- Machinery and Machine Elements
- Mechanical Structures and Stress Analysis
- Automotive Engineering
- Engine Technology
- Aerospace Technology and Astronautics
- Nanotechnology and Microengineering
- Control, Robotics, Mechatronics
- MEMS
- Theoretical and Applied Mechanics
- Dynamical Systems, Control
- Fluid Mechanics
- Engineering Thermodynamics, Heat and Mass Transfer
- Manufacturing
- Precision Engineering, Instrumentation, Measurement
- Materials Engineering
- Tribology and Surface Technology

Indexed by SCOPUS, EI Compendex, and INSPEC.

All books published in the series are evaluated by Web of Science for the Conference Proceedings Citation Index (CPCI).

To submit a proposal for a monograph, please check our Springer Tracts in Mechanical Engineering at https://link.springer.com/bookseries/11693.

Tarek Mabrouki · Habib Sahlaoui · Haifa Sallem ·
Farhat Ghanem · Nourredine Benyahya
Editors

Advances in Additive Manufacturing: Materials, Processes and Applications

Selected Contributions to the 2nd Advances in Additive Manufacturing Conference (AIAM' 2023), May 18–20, 2023, Hammamet, Tunisia

Springer

Editors
Tarek Mabrouki
Applied Mechanics and Engineering Laboratory (LMAI-LR-11ES19)
University of Tunis El Manar (UTM), National Engineering School of Tunis (ENIT)
Tunis, Tunisia

Habib Sahlaoui
Laboratory of Mechanics, Materials and Processes (LMMP) LR-99ES05
University of Tunis (UT), National High School of Engineering of Tunis (ENSIT)
Tunis, Tunisia

Haifa Sallem
HEI-VS, HES SO University of Applied Sciences and Arts Western Switzerland
Powder Technology and Advanced Materials
Sion, Switzerland

Farhat Ghanem
Laboratory of Mechanics, Materials and Processes (LMMP) LR-99ES05
University of Tunis (UT), National High School of Engineering of Tunis (ENSIT)
Tunis, Tunisia

Nourredine Benyahya
Laboratory of Mechanics, Production, and Energy (LMPE)
University of Tunis (UT), National High School of Engineering of Tunis (ENSIT)
Tunis, Tunisia

ISSN 2195-4356 ISSN 2195-4364 (electronic)
Lecture Notes in Mechanical Engineering
ISBN 978-3-031-47783-6 ISBN 978-3-031-47784-3 (eBook)
https://doi.org/10.1007/978-3-031-47784-3

This Springer imprint is published by the registered company Springer Nature Switzerland AG
The registered company address is: Gewerbestrasse 11, 6330 Cham, Switzerland

Paper in this product is recyclable.

Preface

Why Additive Manufacturing?

Additive manufacturing (AM) stands as one of the most remarkable technological revolutions of recent decades. It consists of building, by adding successive layers of material with varying levels of geometrical complexity according to a precise 3D model. Initially confined to 3D printing of plastics, this technology has rapidly evolved, now encompassing a wide array of materials (steels, ceramics, aluminum alloys, titanium alloys, composites, etc.). Indeed, the emergence and development of new AM techniques and processes (SLM, FDM, etc.) lead to make possible the manufacturing of parts and structures from materials previously challenging to cast or mold. In addition, AM, compared to conventional manufacturing methods and processes, offers great flexibility and allows numerous advantages enabling the production of highly intricate geometries with a very high precision. Additionally, it significantly reduces costs, time, and material waste and could deal with different kinds of complexities such as:

- Geometric: All geometric shapes are theoretically possible to fabricate with AM by combining geometric optimizations under constraints (strength, mass, etc.).
- Hierarchical: Production of complex shapes at different scales from micro, meso to the macrostructure.
- Material: The final part or structure could have material anisotropy exhibiting property gradients.
- Functional: Possibility of making assemblies with different mechanisms at the same time and without any future assembly step.

These various advantages have allowed this technology to invade in a few years many sectors and fields such as the automobile, aeronautics, biomedical, civil engineering, and even the manufacturing industries of consumer goods.

This proceeding preface explores the diverse complexities that additive manufacturing can address, presenting its potential to revolutionize industries and open up new possibilities for designers, engineers, and manufacturers alike. Through cutting-edge research and innovation showcased in this volume, we hope to inspire further exploration and harness the full potential of additive manufacturing in shaping a more efficient and sustainable future.

Why a Conference on Advances in Additive Manufacturing?

AM technologies and their applications are in a very rapid evolution and vast expanding stages. The growing interest from industrialists in AM has sparked the enthusiasm of diverse scientific communities across various disciplines. Consequently, these researchers are dedicating their efforts to explore topics related to AM development and the optimization of associated processes. The objectives of their work are diverse, ranging from improving the quality of products obtained, particularly focusing on aspects like morphology, surface roughness, mechanical and physicochemical properties, reliability, accessibility, etc. A great interest is given also to the integration of numerical simulation, machine learning (ML), and artificial intelligence (AI) in AM processes.

The primary purpose of this conference is to provide a platform for researchers, engineers, and industrials to share their latest findings, exchange knowledge, and foster collaboration in the AM field. Potentially, this will help to unlock new horizons and drive innovation, making a lasting impact on the future of manufacturing.

Why This Proceeding?

Through this proceeding, we highlight the importance of collecting a careful selection of the main works presented and perfected over the discussions and enriching exchanges between the participants at the AIAM'2023 conference, which was held in Hammamet, Tunisia, on May 18–20, 2023. This compilation is particularly beneficial by its remarkable diversity:

- Firstly, it incorporates contributions from various fields of research, reflecting the broad and multidisciplinary scope of this conference. Indeed, the participants come from various sectors of activity such as mechanics, materials, robotics, thermal, CAD, civil engineering, and many others. This plurality of disciplines guarantees a fertility of ideas, a fruitful of discussion, and a multitude of perspectives, making it possible to explore innovative and potentially revolutionary solutions.

- Secondly, the geographical diversity of the participants is a key element of the improvement of the content of this proceeding. Researchers and professionals from countries such as Tunisia, France, Switzerland, Canada, Belgium, and Algeria have contributed to this conference, thus giving an international dimension to the scientific and technological advances presented. This openness to the world promotes cultural enrichment, the sharing of knowledge, and the establishment of fruitful international collaborations. All of which has been documented in this proceeding.

- Finally, the participation of the industrials was decisive, and this proceeding bears witness to this. The action fields of the industrialists who supported and participated in the event highlight the relevance and concrete applicability of the presented researches. This promising collaboration between academia and industry is essential to foster the practical implementation of scientific discoveries and to respond effectively to current and future practical challenges.

In summary, this proceeding is invaluable as a compilation of the main scientific and technical contributions from the conference AIAM'2023. It bears witness to the importance of knowledge exchange and interdisciplinary collaboration and highlights the fruitful synergy between academic and industrial in the rapidly developing area of AM. By disseminating these advances to a wider audience, this proceeding will undoubtedly contribute to the advancement of science and technology in the AM area and will open the way for new opportunities and innovative solutions for tomorrow's challenges in this field.

By the Editors

Tarek Mabrouki
AIAM'2023 Conference Chairman

Habib Sahlaoui
AIAM'2023 Scientific Committee Chairman

Contents

Control of the Microstructure and Mechanical Properties of a Super Duplex SAF 2507 Steel Produced by Additive Manufacturing

Maxime Piras[1](✉), Anis Hor[1], and Eric Charkaluk[2]

[1] Institut Clément Ader (ICA), Université de Toulouse, CNRS, ISAE-SUPAERO, UPS, INSA, Mines-Albi, 3 Rue Caroline Aigle, 31400 Toulouse, France
maxime.piras@isae-supaero.fr

[2] Laboratoire de Mécanique Des Solides (LMS), CNRS, UMR 7649, Ecole Polytechnique, Cedex 91128 Palaiseau, France

Abstract. Two-phase duplex stainless steels are widely used in transport industry due to their good fatigue and corrosion properties. Although these steels are investigated when they are produced and/or shaped by conventional processes, few research works have characterized duplex steels obtained by additive processes.

This paper concerns the elaboration and characterization of super duplex steel SAF 2507 obtained by laser powder bed fusion (LPBF). Firstly, the printability of this steel is discussed by analyzing the impact of certain process parameters on the density and microstructure of the material. This as-built microstructure being ferritic and not very ductile, various heat treatments (HT) are, then, applied in order to obtain the typical two-phase microstructure and thus improve the ductility. Afterward, the impact of the different phases (austenite and ferrite) on its mechanical behavior is analyzed.

This work has shown the possibility of obtaining an austenitic-ferritic microstructure close to 50–50 with different grain morphologies with an appropriate post-fabrication TTH. The resulting mechanical properties are satisfactory compared to those of duplex steel obtained by conventional processes.

Keywords: Additive manufacturing · duplex stainless steel · heat treatment · microstructure · mechanical behavior

1 Introduction

Duplex stainless steels (DSS) were developed during the 20th century with the aim of obtaining materials with high mechanical properties and resistance to harsh environments. This alloy has increased resistance to corrosion and fatigue. This allows this material to be used in various industries such as the oil and gas industry, paper mills, chemical transport, storage, naval,... (R.N. Gunn 1997, I. Alvarez-Armas and S. Degallaix 2012).

Many studies in the literature mention DSS produced by conventional processes, but few papers deal with additive manufacturing (AM) processes with these materials. The specificities of AM, such as the greatly different time-temperature profiles with

T. Mabrouki et al. (Eds.): AIAM 2023, LNME, pp. 1–9, 2024.
https://doi.org/10.1007/978-3-031-47784-3_1

conventional processes, lead to major differences in terms of microstructures obtained. Consequently, it is necessary to characterize DSS produced by AM and to compare their properties with conventional processes (T. DebRoy et al. 2018, T. Mukherjee et al. 2016).

Since little information is available in the literature and from AM machine producers (K. Davidson and S. Singamneni 2015, K. Saeidi et al. 2016, S. Papula et al. 2019), it is difficult to define the optimal manufacturing parameters. The various studies also show that it is impossible to obtain a two-phase as-built material when using laser powder bed fusion (LPBF) process. In the literature, the as-built microstructure is almost entirely ferritic despite the different energies and manufacturing strategies (K. Davidson and S. Singamneni 2015, K. Saeidi et al. 2016, S. Papula et al. 2019). It is therefore necessary to carry out a post-fabrication heat treatment (HT) to obtain the duplex material with a suitable ferrite-austenite ratio.

In this paper, an in-depth study of the alloy processability is carried out to optimize the microstructure and to obtain a defect-free material. In order to restore the presence of the austenitic phase, the influence of different HT is studied. The mechanical properties of the optimized microstructures are finally presented.

2 Material and Methods

2.1 Powder

The material used in the study is SAF 2507 (UNS S32750, X2CrNiMoN25-7-4) supplied as powder batches, gas atomized (with nitrogen). Two particle sizes were selected: the first, corresponding to the particle size generally recommended by LPBF manufacturers, i.e. 20–45 μm, and another coarser one, which is generally used for the laser metal deposition (LMD) process, i.e. 45–90 μm. The composition of the two powders is indicated in Table 1. The specific diameters D_{10}, D_{50} et D_{90}, show that the size of the powders does not respect the particle size announced by the manufacturer. This is characterized by the presence of many satellites and non-spherical particles.

Table 1. Chemical composition and particle size of SAF 2507 super duplex steel powders.

Powder	Element (w. %)									Granulometry		
	Fe	Cr	Ni	Mo	Mn	Si	N	Co	Al	D_{10}	D_{50}	D_{90}
Fine (20–45 μm)	Bal	25.0	7,1	3,93	0,8	0,4	0,3	0.1	0.01	21,2	27,4	40,1
Coarse (45–90 μm)	Bal	24.7	6.9	4.04	0.8	0.4	0.29	0.02	0.013	51,8	76	98,1

2.2 LPBF Process

The samples were fabricated using the SLM 125HL machine from the manufacturer SLM Solutions. An inert atmosphere with argon is applied to keep the amount of oxygen below 0.1%. The substrate is preheated to 100 °C. To evaluate the energy delivered to the material during manufacturing, the volume energy density (VED) is generally used (H. Gu et al. 2013) and is calculated as follows:

$$\mathrm{VED} = \frac{P}{v \times h \times t} \tag{1}$$

With P the power in W, v the laser scanning speed in mm.s^{-1}, h the hatch distance in mm and t the layer thickness in mm.

Different VED values have been used in the literature to print this alloy (K. Davidson and S. Singamneni 2015, K. Saeidi et al. 2016, S. Papula et al. 2019). A wide range of VED (from 50 to 233.3 J.mm^{-3}) is firstly used in this study to optimize the energy in relation to the material density. Then, two VEDs were selected (one, medium, at 78.6 J.mm^{-3} and another, higher, at 200 J.mm^{-3}) to analyze the effect of energy on microstructure and mechanical behavior. To obtain the different values of VED, only power and the laser scanning speed are varied. The layer thickness and the hatch distance are therefore constant and equal respectively to 0.05 mm and 0.1 mm.

The lasing strategy can also be modified by imposing or not a rotation angle between the layers. Two strategies are studied: without rotation and with 66° rotation angle.

2.3 Heat Treatments

Naberthem furnaces were used to carry out heat treatments ranging from 1000 °C to 1350 °C. The holding time is also varied between 15 min and 5 h. In order to freeze the microstructure obtained and to avoid the germination of undesirable phases or precipitates, a water quenching was realized.

2.4 Characterization Methods

The X-ray micro-tomography technique was used to assess the samples densities. The Easytom 130 machine from RX Solutions was used with a voltage of 130 V, an exposure time of 4 frame.s^{-1} and an intensity of 61 μA. The obtained voxel size is 6.46 μm. The samples analyzed are cylinders of Ø4 mm and height 10 mm. The volume density of each sample was determined by image analysis using ImageJ software (J. Schindelin 2012).

For the microstructure observations, as-built and heat-treated cubes were cut out to observe the X, Y and Z planes (named according to the normal to the observed face). The X and Y faces are similar for the without rotation lasing strategy and are therefore confused and called XY faces (faces parallel to the building direction). The Z plane is the face normal to the building direction.

Electron backscattered diffraction (EBSD) was used to obtain phase maps and thus determine the proportion of ferrite and austenite phases. The FEG JSM 7100F TTLS LS scanning electron microscope (from JEOL), equipped with the EBSD CMOS Symmetry S2 detector (from Oxford Instruments), was used.

For the tensile tests, the Instron EM 5669 machine was used. The AVE2 video extensometer from Instron has made it possible to carry out the deformation control tests according to the NF EN 2022–001 standard.

Results are compared with a wrought material. It is the SAF 2507 obtained in round bar of 70 mm diameter and solution annealed at 1050 °C during 1 hand water-quenched.

3 SAF 2507 Printability by LPBF Process

3.1 LPBF Material Density

Figure 1 illustrates the relative densities obtained by X-ray micro-tomography analysis of samples built using different VED values for the two powder sizes. Little variation is observed with powder size changing. The results show also that the density of the samples increases with increasing VED to reach 99.9% at 86 J.mm^{-3}. This evolution corresponds to the transition from the pore regime by lack of fusion (LoF) to the convection regime. For high VEDs (from 166 J.mm^{-3}), the density remains high and does not reflect the formation of keyholes generally observed for these energy levels.

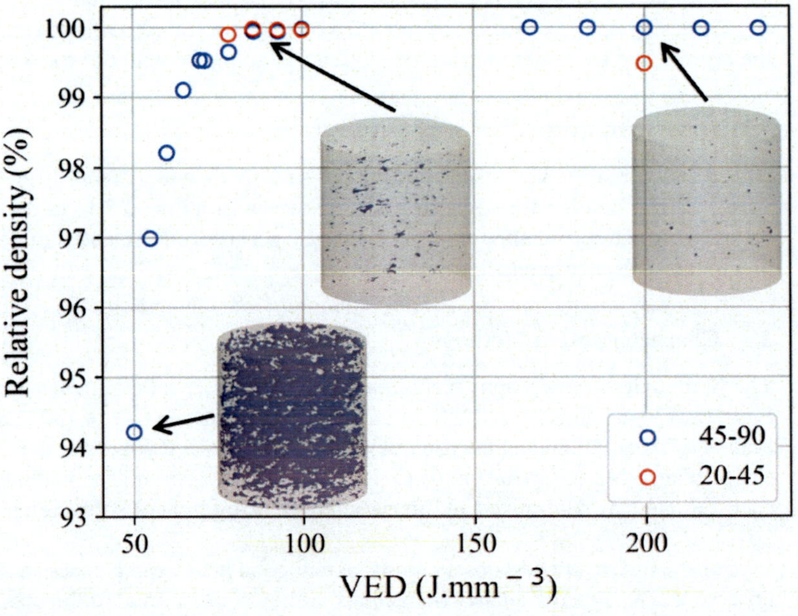

Fig. 1. Relative density evolution as a function of the VED for the two powder granulometries

3.2 As-Built Microstructure

Firstly, due to the very higher cooling rate characterizing the LPBF process (T. DebRoy et al. 2018), the as-built microstructures are a ferrite single-phase. This is observed on the phase maps of Fig. 2, regardless of the VED value used. A maximum of 2.3% austenite is obtained for a higher VED (Fig. 2b). HT are therefore necessary to create the austenite phase and thus obtain a duplex (dual-phase) structure.

Then, the EBSD orientation maps illustrated in Fig. 2 show an absence of epitaxial grain growth, contrary to what is often observed in AM (T. DebRoy et al. 2018).

Finally, energy density (VED) has an influence on the crystallographic texture of the material, since many grains [100] are oriented along the building direction (BD) at VED of 200 J.mm^{-3} while the orientations are more random for a VED of 78.6 J.mm^{-3}.

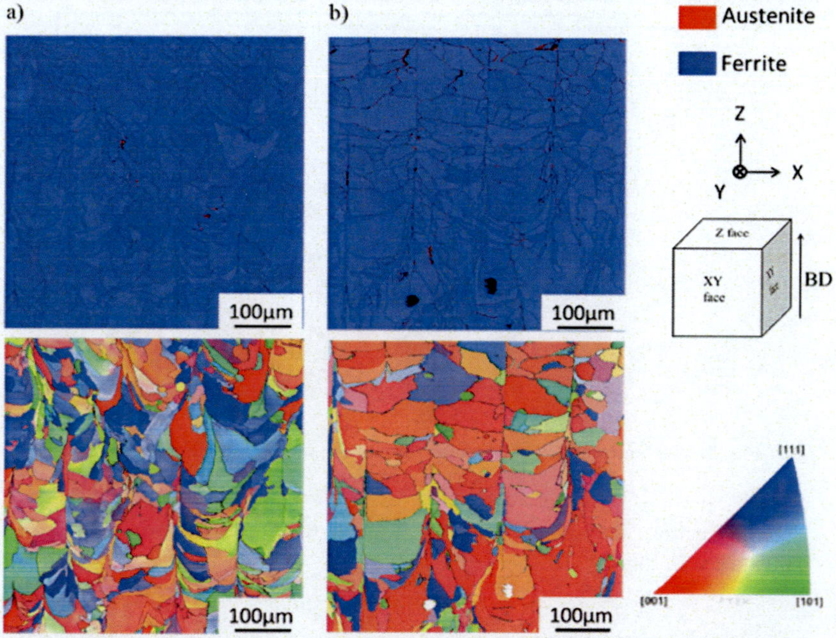

Fig. 2. Orientation and phase maps obtained by EBSD analysis of XY plane for particle size of 45 – 90 µm and using a VED of a) 78,6 J.mm−3 and b) 200 J.mm−3

4 Duplex Microstructure Generation

Different grain morphologies and phase ratio could be obtained by post-fabrication heat treatment.

As shown in Fig. 3, a HT at 1100 °C for 30 min makes it possible to obtain a ferrite/austenite ratio close to 50/50 with a fine microstructure, inherited from the thermal history of the LPBF process.

The increase in holding time and/or temperature make it possible to erase the thermal history and to obtain grain growth, as shown by the HT at 1275 °C for 3 h. A loss of the 50/50 ratio is also observed.

Finally, the dual HT is also performed. It consists of carrying out a first HT at 1275 °C during 3 h followed by a HT at 1100 °C during 1 h. This makes it possible to return to an equivalent proportion between ferrite and austenite and to approach an equiaxed

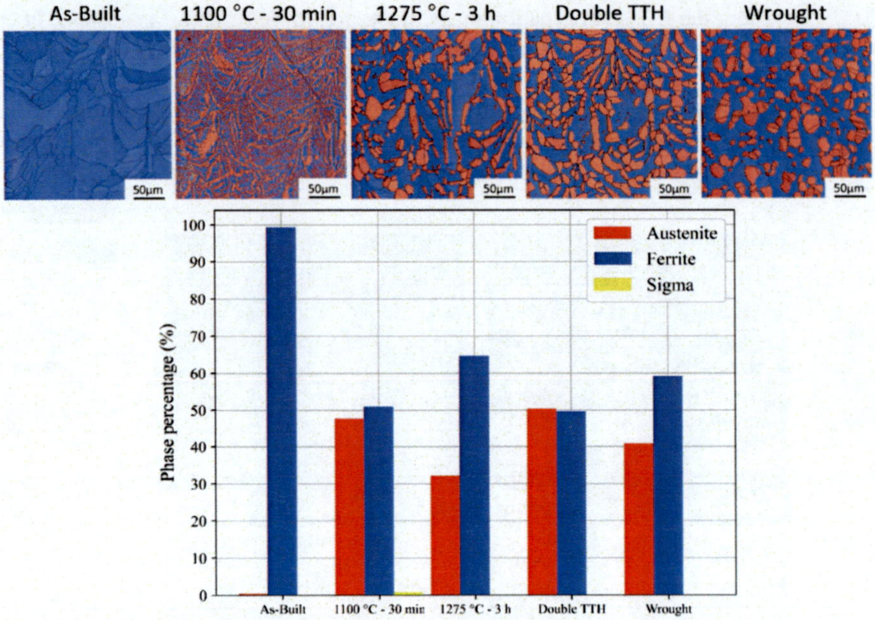

Fig. 3. Austenite- ferrite phase maps and phase ration of the heat-treated microstructures

microstructure. This dual HT made it possible to approach a wrought microstructure as illustrated in Fig. 3.

5 Mechanical Properties

The correlation between the hardness evolution and the austenite-ferrite phase ratio of SAF 2507 according to the applied HT is illustrated in Fig. 4. The hardness is higher for the as-built material which is a ferrite single-phase microstructure. Obtaining a ferrite-austenite ratio close to 50–50 makes it possible to soften the material with a hardness close to wrought microstructure.

The tensile strain-stress curves of as-built and heat-treated microstructure are plotted in Fig. 5. As for hardness, the as-built material shows a particular tensile behavior compared to the other microstructures with a higher yield stress and tensile strength and lower rupture strain. This is due to its almost entirely ferritic composition and the presence of dislocations.

The precipitation of austenite phase by HT makes it possible to obtain behavior close to wrought material. The finer microstructure of HT at 1100 °C during 30 min shows a slightly higher mechanical strength. The difference in phase percentage between the dual HT and that at 1275 °C during 3 h does not lead to different in tensile behavior. However, the strain-stress curves are very close and similar to the wrought material which can be explained by the quasi-equiaxed microstructure morphology.

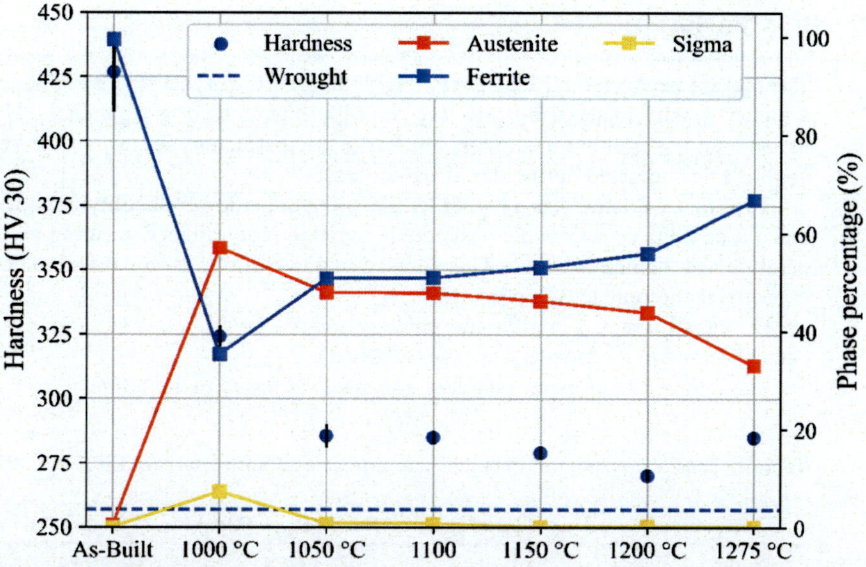

Fig. 4. Evolution of hardness and phase ratio as a function of heat treatment temperatures.

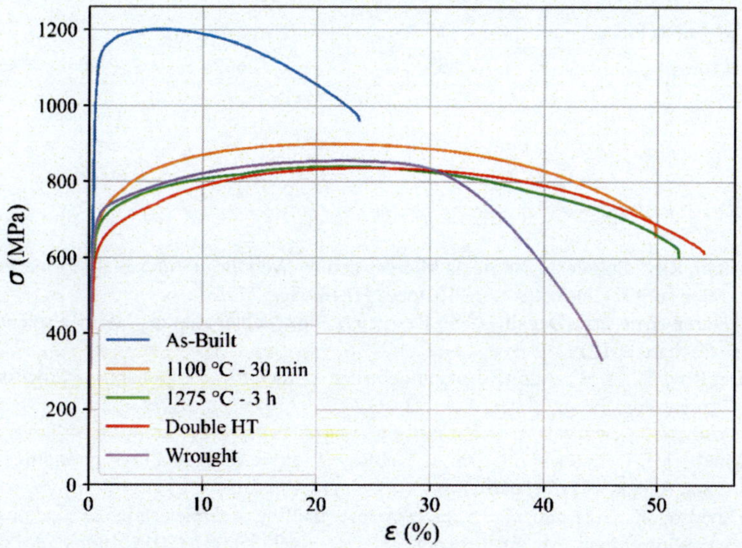

Fig. 5. Tensile strain-stress curves of as-built and heat-treated microstructures of LPBF SAF 2507 DSS

6 Conclusion

The present work on the LPBF SAF 2507 DSS has shown that it is possible to obtain a higher density material by selecting the right manufacturing parameters. However, the as-built material has a single-phase ferrite microstructure because of the very high cooling rates operated during the LPBF process.

This microstructure has a higher tensile strength but lower ductility. Heat treatment make it possible to precipitate austenite in variable proportions depending on the temperature and the holding time. This allowed to find a duplex material and to obtain the expected monotonic tensile properties.

Despite differences in grain morphology and phases ratio, few variations may have been observed in the tensile properties (Table 2).

The fatigue life of these microstructures will be tested in the future.

Table 2. Tensile properties of as built and heat-treated microstructures of LPBF SAF 2507 DSS

	R_m (Mpa)	$R_{e0,2}$ (Mpa)	ε (%)
As-Built	1201	1118	24
1100 °C – 30 min	900	677	50
1275 °C - 3 h	840	640	52
Double HT	837	590	54
Wrought	855	652	45

References

Gunn, R.N., Stainless, D.: Steels Microstructure. Woodhead Publishing, Properties and Applications (1997). https://doi.org/10.1533/9781845698775.24

Alvarez-Armas, I., Degallaix, S.: Les aciers inoxydables duplex. Hermès science publications, Cachan (2012)

DebRoy, T., et al.: Additive manufacturing of metallic components – Process, structure and properties. Prog. Mater. Sci. **92**, 112–224 (2018). https://doi.org/10.1016/j.pmatsci.2017.10.001

Mukherjee, T., Zuback, J.S., De, A., DebRoy, T.: Printability of alloys for additive manufacturing. Sci. Rep. **6**, 19717 (2016). https://doi.org/10.1038/srep19717

Davidson, K., Singamneni, S.: Selective laser melting of duplex stainless steel powders: an investigation. Mater. Manuf. Process. **31**, 150930095558007 (2015). https://doi.org/10.1080/10426914.2015.1090605

Saeidi, K., Kevetkova, L., Lofaj, F., Shen, Z.: Novel ferritic stainless steel formed by laser melting from duplex stainless steel powder with advanced mechanical properties and high ductility. Mater. Sci. Eng. A **665**, 59–65 (2016). https://doi.org/10.1016/j.msea.2016.04.027

Papula, S., et al.: Selective Laser Melting of Duplex Stainless Steel 2205: Effect of Post-Processing Heat Treatment on Microstructure. Mechanical Properties, and Corrosion Resistance, Materials. **12**, 2468 (2019). https://doi.org/10.3390/ma12152468

Gu H., Gong, H., Pal, D., Rafi, H., Starr, T., Stucker, B.: Influences of Energy Density on Porosity and Microstructure of Selective Laser Melted 17–4PH Stainless Steel (2013)
Schindelin, J., et al.: A. Cardona, Fiji: an open-source platform for biological-image analysis. Nat. Methods **9**, 676–682 (2012). https://doi.org/10.1038/nmeth.2019

Microstructure and Mechanical Properties of Hybrid LPBF-DED Inconel 625

Noémie Martin[1,2](✉), Anis Hor[1], Etienne Copin[1], Philippe Lours[1], and Léon Ratsifandrihana[2]

[1] Institut Clément Ader (ICA), Université de Toulouse, CNRS, IMT Mines Albi, INSA, ISAE-SUPAERO, UPS, 3 Rue Caroline Aigle, 31400 Toulouse, France
nmartin@ethz.ch

[2] SEGULA Engineering, Immeuble EQUINOX - Bâtiment I, 24 Boulevard Déodat de Séverac, 31770 Colomiers, France

Abstract. The hybridization of Laser Powder Bed Fusion and Direct Energy Deposition could increase the application range of additive manufacturing by benefiting from the resolution of LPBF and the flexibility of DED. However, the microstructures and mechanical properties obtained by these processes are very different. This causes the hybrid parts to be very heterogeneous. The mechanical behavior of hybrid LPBF-DED Inconel 625 parts are investigated under static and cyclic loading. Both as-built and heat-treated Inconel 625 are studied to investigate the role of the specific as-built microstructure. Tensile tests are performed using Digital Image Correlation, which allows the global behavior to be explained by the local behavior of the samples. Hybrid samples are also tested under high cycle fatigue, and fractographic observations permit to determine the initiation sites and mechanisms. The DED is found to be the limitation in both the static and the fatigue strengths of hybrid samples. Hybrid samples perform as well as or better than mono-process DED. In static loading, the local behavior of the hybrid samples and the yield tensile strength gap between LPBF and DED determine the global behavior. The heat-treatment is successful in reducing this gap, and improving the global behavior of the hybrid samples, especially the total elongation. In fatigue, the defects are preponderant compared to the mechanical heterogeneity of hybrid samples.

Keywords: additive manufacturing · hybrid · microstructural heterogeneity · heat treatment · fatigue behavior

1 Introduction

Additive manufacturing of metallic alloys is attractive for many industrial fields, from the aeronautics, to the medical or even jewelry (Buchanan and Gardner 2019). Initially used for prototyping (Wohlers and Caffrey 2018), it is now used for repair and adding features on existing parts, or as the main production tool for certain parts. This variety of applications pulled the development of a large panel of technologies, each with their own specific characteristics (Gao et al. 2015).

T. Mabrouki et al. (Eds.): AIAM 2023, LNME, pp. 10–18, 2024.
https://doi.org/10.1007/978-3-031-47784-3_2

Among the most established, Laser-Powder Bed Fusion (LPBF) provides a high geometrical resolution and satisfactory mechanical properties, but with high costs, low production rate and a specific built plate as substrate. Direct Energy Deposition (DED) achieves higher deposition rates, and can be used on any substrate geometry (Gibson et al. 2015). The hybridization of those two processes and the combination of their strengths could increase the application cases available to additive manufacturing.

However, the microstructures obtained by LPBF and DED and the resulting mechanical properties are very different. Studies on hybrid LPBF-DED with Ti6AV have been conducted and all concluded that the tensile strength of such structures are limited by the DED section of the samples (Graf et al. 2015; Liu et al. 2016; Qin et al. 2019). Similar conclusions were drawn on steel (Oh et al. 2019), with an emphasis on the importance of the LPFB-DED interface. Cracks, pores or other macro defects can significantly decrease the tensile properties of the hybrid part, such as yield strength and elongation. (Godec et al. 2021) studied Inconel 718 and justified the difference of mechanical properties between LPBF and DED by the grain size, texture, and dislocation density, investigated by the comparison with heat-treated hybrid samples. These hypotheses have been confirmed on Inconel 625 (Martin et al. 2022).

Studies of the mechanical properties under quasi-static loading of hybrid LPBF-DED samples exist in the literature, but information about the fatigue behavior is lacking. Hybrid 316L parts displayed lower fatigue properties than mono-process parts in the investigation by (Balit et al. 2020), but the cause of this deterioration was not identified.

In the following investigation, the tensile and fatigue properties of hybrid LPBF-DED Inconel 625 and the local behavior of the samples are described and analyzed. The role and impact of the LPBF-DED interface and the difference of properties between the LPBF and the DED are studied experimentally. Tensile tests instrumented with local strain measurements give access to the global and local static behaviors of the material. Hybrid samples are also characterized in high-cycle fatigue. Finally, the comparison of as-built and heat-treated (homogenized) samples highlight the impact of the difference of properties between LPBF and DED.

2 Materials and Method

2.1 Fabrication of Samples

Cylinders of 12 mm and 20 mm were printed on a SLM Solution 125HL LPBF machine. Then, these cylinders were removed from the chamber, cleaned, and completed up to 100 mm height by DED in a BeAM-AddUp Modulo 400. The chemical composition of the powders used are defined by the standard UNS06625, and the granulometries are given on the Table 1 along with the process parameters (recommended by the machine's suppliers).

The annealing treatment used is 4 h at 1150 °C, ended with a water quench. This treatment allows a full recrystallisation and homogenization of the LPBF and DED Inconel 625 obtained by these parameters (Martin et al. 2022).

The cylinders were then machined in two different geometries: flat samples for tensile tests and cylindrical samples for fatigue tests. The fatigue samples were also manually polished to obtain a roughness below 5 μm. For each sample, the interface between LPBF and DED is located at the center of the gauge length.

Table 1. Parameters for LPBF and DED fabrications.

LPBF		DED	
Power	275 W	Power	900 W
Speed	760 mm/s	**Speed**	13.3 mm/s
Layer	50 μm	**Flow**	9.5 g/min
Powder	21–54 μm	**Powder**	47–88 μm

2.2 Mechanical Tests

Vickers micro-hardness was measured by indentation using a 136° pyramid loaded during 10s at 3 kg. 6 × 10 matrices with 1 mm millimeter increment between columns and 0.5 mm between lines were performed placed so that the interface between LPBF and DED was at mid-height.

As specified by the NF EN 2002–001 standard, the tensile tests were controlled by strain rate with an optic extensometer focused on the length of the working section of the samples. In addition, the samples were speckle-painted for digital image correlation (DIC) analysis. Both cameras took images with an acquisition rate of 2 Hz. The software VIC3D was used for the analysis.

The fatigue experiments were controlled via stress, on sinusoidal signal of 15 Hz and loading ratio of 0.1. Three types of samples were tested for comparison: made by LPBF, by DED and by hybridization.

2.3 Microstructural Characterization

Optical and scanning electron microscopies were used on mechanically polished down to mirror finish, and etched samples. The etchant used was Aqua Regia solution (30%$_{vol}$ HNO3 – 70%$_{vol}$ HCl).

3 Results

3.1 Hybrid LPBF-DED Inconel 625: Microstructure

The microstructure of the hybrid samples around the interface is given on Fig. 1. In both the LPBF and the DED sections, the ex-melt pools are visible, crossed by grains that are themselves made of dendrites structures. All these features are finer in the LPBF by a factor 10 compared to the DED (for example, the LPBF dendrites are 1 μm wide, against 10 μm for the DED in average). These microstructures, out of equilibrium, are commonly observed in additively manufactured Inconel 625 (Marchese et al. 2017). Finally, the grain morphology and size are homogeneous in LPBF section and heterogeneous in DED section (Fig. 1c).

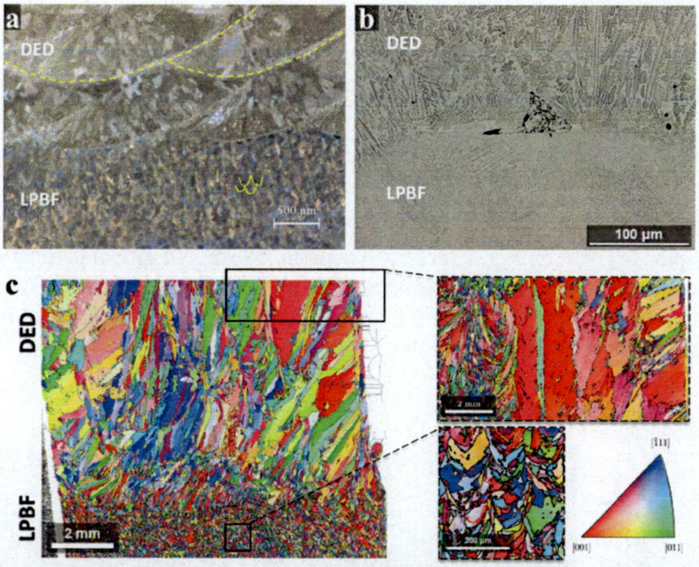

Fig. 1. Microstructure of the as-built hybrid samples around the LPBF-DED interface: a) optical, b) SEM observations and c) EBSD grain orientation maps.

The interface does not present significant defects, other than occasional lack of fusion (LoF). No cracks, nor specific crystallographic features have been observed.

After the heat treatment, the hybrid samples are entirely recrystallized, with equiaxial grains and weak texture. The dendritic structures are fully erased. The grain size remains higher in the DED section by roughly a factor 10, which is the only remaining difference be-tween LPBF and DED heat-treated microstructure.

3.2 Micro-hardness

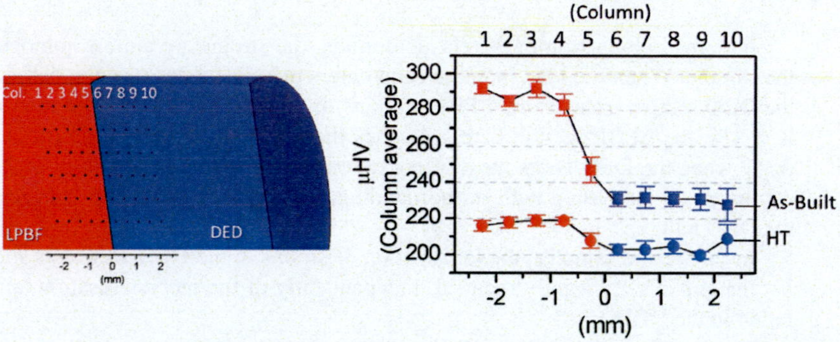

Fig. 2. Microhardness of hybrid as-built and heat-treated samples: indents matrix & results.

The micro-hardness measurements across the interface reveal a neat evolution from 290 μHV in the LPBF to 230 μHV in the DED section. The average of each column of the indentation matrix is given on Fig. 2. This is consistent with the difference in microstructure observed, finer in the LPBF than in the DED Inconel 625.

3.3 Tensile Behavior

The evolution of the fields of strain during the tensile tests for a as-built and a treated samples is given on Fig. 3. For each sample, 6 levels of global elongation ($\varepsilon_{\text{global}}\%$) is given by the optical video extensometer. The mean strain of the LPBF and DED sections, calculated from the DIC results, are given in white on each picture. Finally, the local strain is given by the color scales.

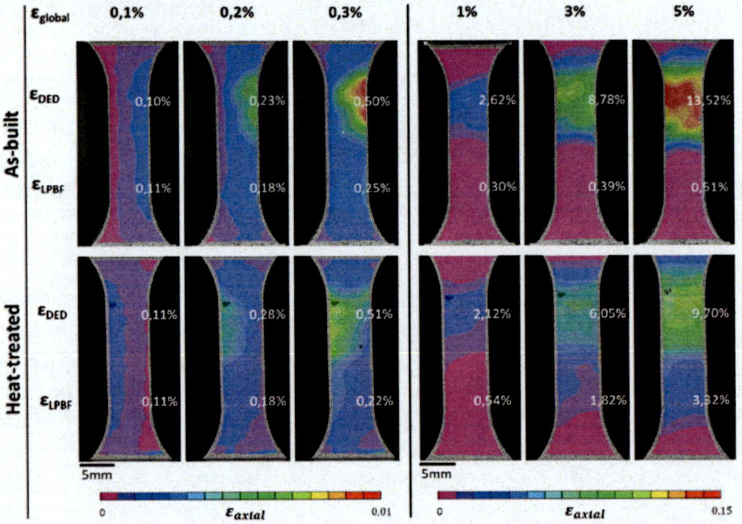

Fig. 3. Evolution of the strain field during a tensile test for hybrid samples (as-built and heat-treated).

For small elongation in the elastic domain, the strains are almost homogeneous in the sample. When the load increases and enters the plastic domain, the deformation of the DED section exceeds the LPBF sections. In the as-built state, when the total strain is at 1%, the LPBF section barely reaches the plastic domains with a mean strain of 0.3% when the DED bears most of the deformation with a deformation of 2.6%. The gap keeps on increasing with further loading, with a clear localization of the strains in the DED half.

In the treated samples, the localization of strains in the DED section is still visible, but the gap is significantly reduced. Independently of the recrystallisation, all samples failed in the DED.

The local behavior of the LPBF and DED sections of the hybrid samples have been traced and compared to LPBF and DED samples on Fig. 4. The local behaviors of hybrid

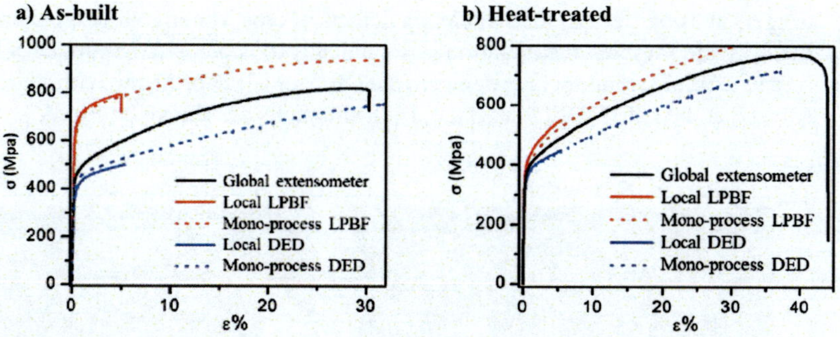

Fig. 4. Comparison of the local behavior and global behavior of hybrid samples a) as-built b) heat-treated.

samples were identical to the respective mono-process for both as-built and heat-treated states. The hybrid samples show a mechanical resistance in between LPBF and DED properties, but with a lower global elongation.

3.4 Tensile Behavior

Figure 5 a and b are the S-N plots for LPBF, DED and hybrid samples as-built and heat-treated respectively. The load level is represented by the maximum stress level σ_{max}. In as-built state, the fatigue lives of LPBF and DED samples are similar for high stress levels for which the main factor is crack propagation. The hybrids show a slightly inferior fatigue resistance. For lower loadings, where initiation becomes preponderant, LPBF samples show significantly reduced fatigue lives compared to DED and hybrid samples that have similar fatigue lives. In the heat-treated state, all samples show similar fatigue resistance.

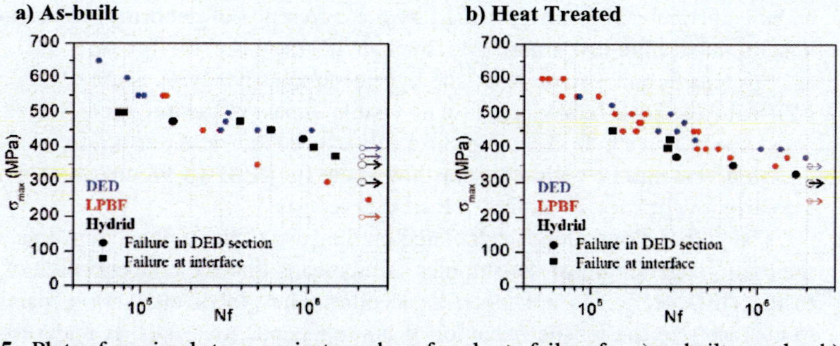

Fig. 5. Plots of maximal stress against number of cycles to failure for a) as-built samples, b) heat treated samples. Red: LPBF, blue: DED, black: hybrid specimen. (R0.1, 15 Hz).

The crack initiation in the LPBF samples was systematically localized on a lack of fusion close to the surface of the sample. For the samples made by DED, the initiation triggers at the surface of the sample, on defects of type Marangoni or on microstructure. In the hybrid samples, the initiation systematically took place at the surface of the sample on a defect: a Marangoni type defect in the DED, or a lack of fusion with a circular shape situated at the LPBF-DED interface. Both situations are shown on Fig. 6.

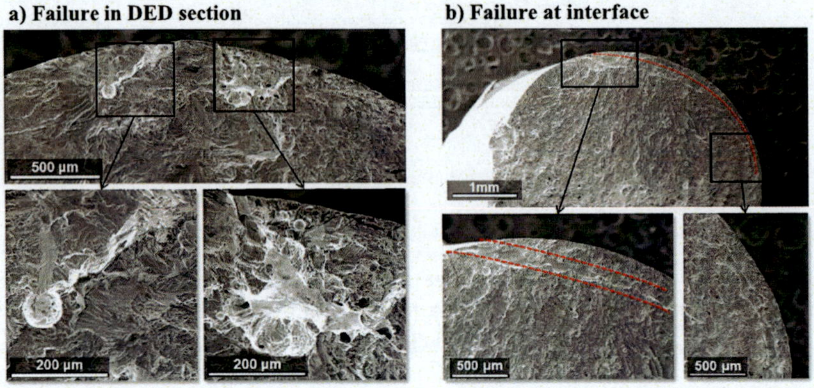

Fig. 6. Initiation sites for fatigue crack on hybrid samples a) in DED section b) at interface.

4 Discussion

4.1 Static Properties and Microstructure

The micro-hardness measurements translate what is observed on the microstructure, with a gap between as-built PLBF and DED. After the microstructures are heat treated and recrystallized, the gap is significantly decreased. Yet, the grain size difference between as-built and heat-treated LPBF vs DED is similar. The difference in hardness of the as-built Inconel 625 is also justified by the difference of dendritic size and dislocation density rather than just grain size (Hu et al. 2018).

For tensile quasi-static tests, the hybrid samples behavior is the compilation of the LPBF and the DED behaviors, with no visible impact of the interface. Indeed, the global stress applied on a hybrid sample can easily exceed the yield tensile stress of DED (YTS 433MPa), but still be within the elastic domain for LPBF (YTS 667MPa). This explains the early localization of strains in the DED section.

The global elongation is calculated on the base of the video extensometer, which is the total gauge length of the samples. This means that the concentration of the strains in the DED section is not taken in account. The global elongation, parameter used to characterize the tensile behavior of homogeneous materials, is underestimated and irrelevant in this case.

In heat-treated samples, the LPBF and DED sections have less difference in properties (respective YTS 363 MPa and 360 MPa and UTS 853 MPa and 768 MPa). Hence the localization phenomenon is reduced. In all cases, the hybrid samples displayed yield and ultimate tensile strengths comprised between the LPBF and DED properties.

4.2 Fatigue Properties: Defects

The hybrid samples all failed on a defect: Marangoni pores or a lack of fusion at the interface. Both these defects are linked to the control of the DED process. The lack of fusion observed at the LPBF-DED interface has a circular shape, corresponding to the deposition strategy used. The first layer is deposited on a called substrate, which explains that this defect is not observed on higher layers, or in mono-process DED samples.

In the as-built state, the LPBF samples showed systematic initiation on LoF. These are no longer observed in hybrid samples because they have been machined out of a cylinders, when the mono-process LPBF have been printed near net-shape and then polished.

After the heat-treatment, it is shown that the LoF have no impact on the fatigue lives of the LPBF samples (Fig. 5.b). The damage tolerance of Inconel 625 is significantly increased. All heat-treated samples show endurance limits close to their YTS (363 MPa).

5 Conclusion

This experimental paper demonstrates the feasibility of LPBF-DED hybridization. DED is confirmed in being the limiting factor, but further understanding is drawn:

- Sections of hybrid samples have the same behavior as mono-process LPBF and DED
- Static behavior of hybrid sample is determined by local behavior, in particular the lower DED's YTS.
- On the contrary, fatigue properties of hybrid samples are mainly impacted by surface defects

References

Balit, Y., Joly, L.R., Szmytka, F., Durbecq, S., Charkaluk, E., Constantinescu, A.: Self-heating behavior during cyclic loadings of 316L stainless steel specimens manufactured or repaired by Directed Energy Deposition. Mater. Sci. Eng. A **786**, 139476 (2020). https://doi.org/10.1016/j.msea.2020.139476

Buchanan, C., Gardner, L.: Metal 3D printing in construction: A review of methods, research, applications, opportunities and challenges. Eng. Struct. **180**, 332–348 (2019). https://doi.org/10.1016/j.engstruct.2018.11.045

Gao, W., et al.: The status, challenges, and future of additive manufacturing in engineering. CAD Comput. Aided Design **69**, 65–89 (2015). https://doi.org/10.1016/j.cad.2015.04.001

Gibson, I., Rosen, D., Stucker, B.: Additive Manufacturing Technologies. Springer, US (2015)

Godec, M., et al.: Hybrid additive manufacturing of Inconel 718 for future space applications. Mater Charact 172 (2021). doi:https://doi.org/10.1016/j.matchar.2020.110842

Graf, B., Schuch, M., Kersting, R., Gumenyuk, A., Rethmeier, M.: Additive process chain using selective laser melting and laser metal deposition. In: Lasers in Manufacturing 2015, p 59 (2015)

Hu, Y.L., et al.: Effect of solution heat treatment on the microstructure and mechanical properties of Inconel 625 superalloy fabricated by laser solid forming. J. Alloys Compd. **767**, 330–344 (2018). https://doi.org/10.1016/j.jallcom.2018.07.087

Liu, Q., et al.: Microstructure and mechanical properties of LMD-SLM hybrid forming Ti6Al4V alloy. Mater. Sci. Eng. A **660**, 24–33 (2016). https://doi.org/10.1016/j.msea.2016.02.069

Marchese, G., et al.: Characterization and comparison of inconel 625 processed by selective laser melting and laser metal deposition. Adv. Eng. Mater. **19**, 1–9 (2017). https://doi.org/10.1002/adem.201600635

Martin, N., Hor, A., Copin, E., Lours, P., Ratsifandrihana, L.: Correlation between microstructure heterogeneity and multi-scale mechanical behavior of hybrid LPBF-DED Inconel 625. J. Mater Process Technol. 303 (2022). https://doi.org/10.1016/j.jmatprotec.2022.117542

Oh, W.J., Lee, W.J., Kim, M.S., Jeon, J.B., Shim, D.S.: Repairing additive-manufactured 316L stainless steel using direct energy deposition. Opt. Laser Technol. **117**, 6–17 (2019). https://doi.org/10.1016/j.optlastec.2019.04.012

Qin, L.Y., et al.: Microstructure homogenizations of Ti-6Al-4V alloy manufactured by hybrid selective laser melting and laser deposition manufacturing. Mater. Sci. Eng. A **759**, 404–414 (2019). https://doi.org/10.1016/j.msea.2019.05.049

Wohlers, T., Caffrey, T.: Wohlers Report 2018 (2018)

Manufacturability of CoCrFeNiMnzAlxCuy High Entropy Alloy by Laser Powder Bed Fusion

Eric Barth(✉) and Anis Hor

Institut Clément Ader (ICA), Université de Toulouse, CNRS, ISAE-SUPAERO, UPS, INSA, Mines-Albi, 3 Rue Caroline Aigle, 31400 Toulouse, France
{eric.barth,anis.hor}@isae-supaero.fr

Abstract. Six pre-alloyed high entropy alloys powders from the $CoCrFeNiMn_zAl_xCu_y$ family were used to manufacture samples through Laser Powder Bed Fusion technology. These samples were used to assess the manufacturability, microstructure and mechanical properties of the developed alloys. It was found that increasing the Al concentration promoted the formation of a BCC/B2 phase, increasing the hardness and having a direct positive impact on the tensile properties. However, adding more aluminium than what can be found in the $Al_{0.5}CoCrFeNi$ alloy led to extensive manufacturability issues that hindered the tensile properties. On the other hand, the Cu concentration did not appear to yield any significant impact on the microstructures and mechanical properties of the alloys. The considered additively manufactured alloys presented an overall higher hardness than their conventionally manufactured counterparts, but a lower ductility. Thermal treatment led to the precipitation of a σ phase in Al-containing HEAs, and of a secondary FCC phase in Cu-containing alloys. Finally, this paper proves the potential of the $CoCrFeNiAl_x$ alloy family compared to single-phase stainless steels such as 316L and two-phase stainless steels such as SAF 2507.

Keywords: Additive manufacturing · Mechanical properties · microstructure · High entropy alloy

1 Introduction

High Entropy Alloys (HEAs) are a mixture of five or more alloys in equimolar or nearly equimolar proportions [1]. They were first introduced by Cantor et al. [2] in 2004. Since then, these alloys have been extensively studied in the literature [1, 3, 4] as they represent a promising alloying concept for next generation materials. The most studied HEA is the Cantor alloy [2, 5], an equimolar mixture of Co, Cr, Fe, Mn and Ni. However, its properties remain far from being outstanding, as they are close to what can be achieved by a conventional 316L stainless steel [6]. This spurred the publication of various studies aiming at doping the properties of the Cantor alloy by adding new elements [7]. However, these efforts remain limited within the sphere of additively manufactured HEAs.

Additive Manufacturing (AM) presents various characteristics that could greatly enhance the manufacturability and properties of HEAs. Their high-density [1] could

T. Mabrouki et al. (Eds.): AIAM 2023, LNME, pp. 19–27, 2024.
https://doi.org/10.1007/978-3-031-47784-3_3

be mitigated by optimized shapes allowed by AM processes. Moreover, it has been found that AM-produced materials exhibit better mechanical properties compared to their conventionally manufactured counterparts [6]. So far, the most commonly used AM technique is Laser Powder Bed Fusion (LPBF), due to its high level of maturity, precision and efficiency.

Cantor alloy, or CoCrFeMnNi alloy was previously manufactured with an LPBF process by Zhu et al. [8]. As with conventional processes, the obtained properties remain similar or lower to what was observed for LPBF conventional alloys made of the same elements. Some efforts were made to further increase the properties of the LPBF-manufactured Cantor alloy by adding Al and Cu in various proportions. It was shown that substituting Mn for various proportions of Al allowed obtaining better hardness and tensile properties due to the precipitation of a hard BCC phase and to the presence of dense dislocation structures [9]. However, going above a critical Al concentration leads to a very brittle behaviour of the alloy and to extensive printability issues, such as cracking due to thermal residual stress [10]. Alternatively, substituting Mn for Cu while remaining equimolar was shown to lead to a highly ductile FCC-phased alloy [11].

2 Materials and Methods

2.1 Powders

CoCrFeMnNi, $Al_{0.3}$CoCrFeNi, $Al_{0.5}$CoCrFeNi, AlCoCrFeNi, AlCuCoCrFeNi and CuCoCrFeNi HEAs were additively manufactured using gas-atomised pre-alloyed powders. The compositions of each powder were analysed by EDX and exhibited the right proportions. The granulometry given by the supplier are 45-90μm for all the powders. The granulometry was also determined using laser diffraction and the obtained results match the values given by the manufacturer. The powders' chemical homogeneity and morphology were analysed using EDX analysis and SEM imaging respectively. All powders present an overall spherical shape, short of a number of satellites and oblong shaped powder grains. Likewise, all powders have good chemical homogeneity, even if some Fe-Rich and Cr-rich grains can be observed in the Al-containing powders.

2.2 Sample Building

Using the aforementioned powders, test samples were built with a SLM 125 HL additive manufacturing machine.

The original plan was to try and apply a single set of building parameters to all the considered alloys, as conducting a parametric optimisation for each alloy proved to be too time-consuming. However, manufacturing issues emerged, and led to a modification of the laser power for each individual alloy. It was then decided to set the hatch distance (h) at 0.1, the layer thickness (t) at 0.05 and the laser velocity at 1000 mm/s (except for Al0.3CoCrFeNi, that used $V = 800$ mm/s and $h = 0.05$). The laser power (P) was then varied between 200 W and 300 W until the build-job was able to be successfully completed. The samples were printed both with a vertical and horizontal build-direction, except for Al0.3CoCrFeNi, whose samples were retrieved from a previous build batch where no vertically printed samples were manufactured. The specimens were cut from the baseplate using an Electric Discharge Wire Cutting machine (EDM).

2.3 Heat Treatment

Samples were heat-treated at temperatures of 600 °C, 800 °C and 1100 °C for a duration of 3 h, followed by water-quenching.

2.4 Sample Analysis

The samples' micro-hardness was analysed with a Vickers indent and a 1kg load applied for 20 s. At least 5 measurement per sample orientation were performed. Tensile tests with 6 × 2.5 mm cross-section specimens (Fig. 1) were performed on an Electromechanical Instron 5900R testing system with at least three tensile tests per build direction and per alloy. Electron backscatter diffraction (EBSD) mappings were performed with a JEOL 7100F SEM. The X-ray diffraction (XRD) measurements for phase characterization were performed with a Bruker's X-Ray diffractometer, using a Cu Kα radiation. The 2θ angle ranged from 28.5° to 105°.

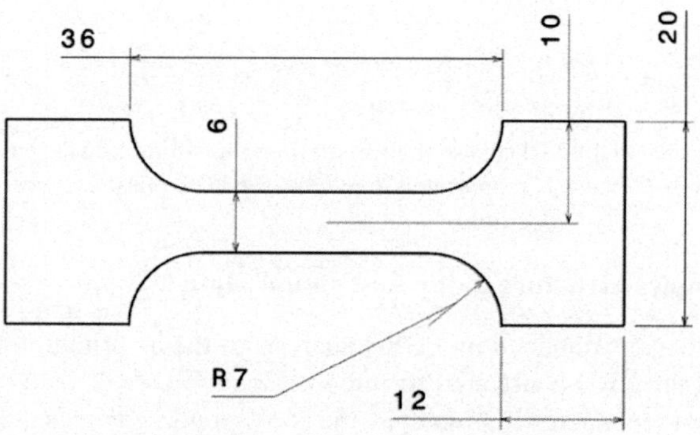

Fig. 1. Blueprint for the used tensile specimens

3 Results

3.1 Chemical Composition

The alloys in this study are composed of elements with melting points ranging from 660 °C for Al to 1857 °C for Co. These differences can complicate the selection of processing parameters, as using a laser power high enough to melt the high melting point elements could evaporate the low-melting point ones. One way of verifying if these parameters are adapted is to check the composition of the sample and compare it with the powder's composition. Here, no major discrepancy was observed. It was noted however that the Al concentration is systematically lower in the sample as it is in the used powder, meaning that a fraction of this element was vaporized during the process.

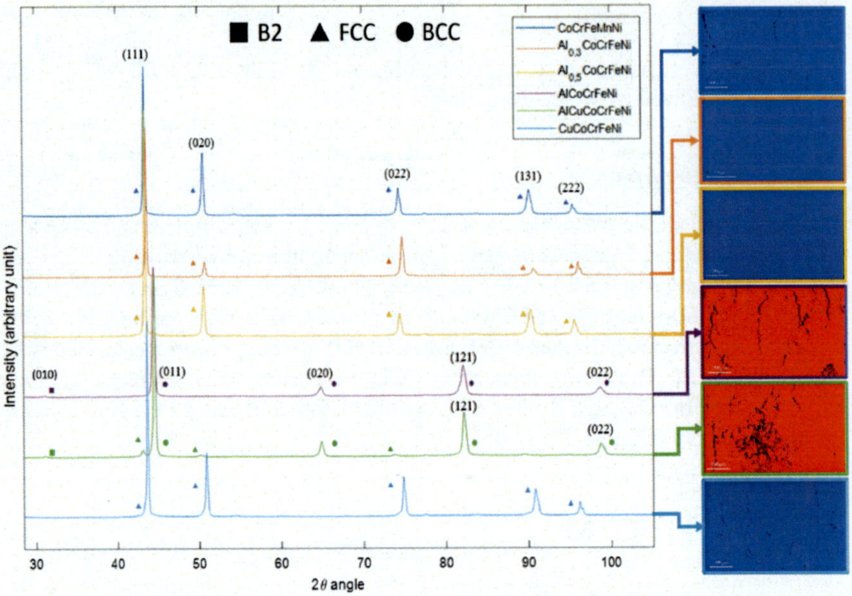

Fig. 2. Results of the XRD phase analysis and corresponding EBSD phase maps for the considered HEA. The BCC phase is represented in red and the FCC phase is represented in blue.

3.2 Samples Structure in the As-Printed State

The phase maps obtained by EBSD analysis of the as-printed samples, as well was the results of the XRD phase analysis are visible in Fig. 2.

We can see that for each sample, the corresponding map is monochromatic, meaning that a single phase was detected per alloy. All alloys are pure FCC, except for AlCoCrFeNi and AlCuCoCrFeNi, which are pure BCC. However, as visible on the XRD analysis, $Al_{0.5}CoCrFeNi$ and AlCuCoCrFeNi both exhibit a dual-phased composition, with small proportions of BCC and FCC, respectively.

It can be seen that CoCrFeMnNi, $Al_{0.3}CoCrFeNi$ and $Al_{0.5}CoCrFeNi$ present an overall dense structure, with a low concentration of pores and cracks, even without optimized processing parameters. It is to be noted however that the CoCrFeMnNi sample exhibits a high concentration of micro-cracks.

AlCoCrFeNi, AlCuCoCrFeNi and CuCoCrFeNi present a very high concentration of cracks and porosities. These cracks appear to be aligned with the build direction, and mostly located at the centre of melt-pools, propagating along the grain boundaries.

Using the same EBSD analysis, the average grain size was determined. It appeared that AlCoCrFeNi and AlCuCoCrFeNi had the finest microstructure, with an average grain size of 20.3μm and 17.6μm, respectively.

3.3 Mechanical Properties in the As-Printed State

The obtained hardnesses and tensile properties of the HEA samples are summarized in Fig. 3(a). The reference Cantor alloy (CoCrFeMnNi) average hardness was measured to be 226 HV. By removing Mn and increasing the Al concentration to obtain the Al0.5CoCrFeNi alloy leads to a 263 HV micro-hardness, a 16% increase compared to the Cantor reference. Augmenting Al concentration until an equimolar concentration is reached (AlCoCrFeNi) more than doubles the micro-hardness compared to the reference Cantor alloy, with a measured average hardness of 525 HV. Adding Cu in equimolar proportions (AlCuCoCrFeNi) leads to a slightly higher average micro-hardness of 550HV.

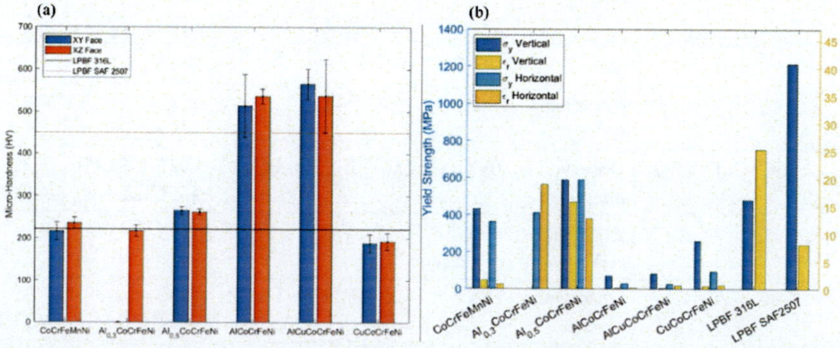

Fig. 3. (a) Microhardness of the samples measured on the XY and XZ faces, with corresponding error bars. The hardnesses of the 316L and SAF 2507 are indicated for comparison purposes. (b) Yield strength elongation at failure of the samples, measured with vertically and horizontally printed samples

By removing Al from the alloy to obtain an equimolar CuCoCrFeNi, we get a measured micro-hardness of 191 HV, a 15% decrease from the reference Cantor alloy.

The tensile tests results of the HEAs are visible on Fig. 3(b). To the exception of Al0.3CoCrFeNi and Al0.5CoCrFeNi, all alloys show a brittle or quasi-brittle tensile behaviour, with very little ductility. Removing Mn and adding Al in the alloy initially increases its tensile strengths, however, going beyond the Al concentration of Al0.5CoCrFeNi leads to a major drop in tensile properties. This is mainly due to the high concentration of pores and microcracks present in the samples. This also applies to AlCuCoCrFeNi. The vertically printed CuCoCrFeNi alloy exhibits a much higher yield and ultimate tensile strength than the horizontally printed ones. However, its tensile performance still remains below the ones of the reference Cantor alloy.

3.4 Effect of Thermal Treatment

The results of the heat-treatment, in terms of observed phases and measured microhardness, are reported in Table 1.

Table 1. Observed phases and microhardness in the AHE samples in the as-printed samples and after heat-treatment at 600 °C, 800 °C and 1100 °C for 3h followed by water-quenching

Alloy		As-printed	600 °C 3h	800 °C 3h	1100 °C 3h
CoCrFeMnNi	Observed phases	FCC	FCC	FCC	FCC
	µhardness (HV1)	226	248	229	212
$Al_{0.5}$CoCrFeNi	Observed phases	FCC + BCC	FCC + BCC	FCC + BCC + σ	FCC + BCC
	µhardness (HV1)	273	355	417	266
AlCoCrFeNi	Observed phases	BCC	BCC	BCC + FCC + σ	BCC + FCC
	µhardness (HV1)	465	545	492	318
AlCuCoCrFeNi	Observed phases	BCC + FCC	BCC + FCC	FCC + BCC + FCC2 + σ	FCC + BCC + FCC2
	µhardness (HV1)	466	473	357	300
CuCoCrFeNi	Observed phases	FCC	FCC + FCC2	FCC + FCC2	FCC + FCC2
	µhardness (HV1)	183	207	197	197

Except for the reference Cantor alloy, multiple new phases appear in all of the AHE. The small amount of BCC and FCC phase observed in $Al_{0.5}$CoCrFeNi and AlCuCoCrFeNi respectively increases with temperature. A secondary FCC phase appears at higher treatment temperature for the AlCoCrFeNi.

For all Al-containing alloys, a σ phase is observed for the 800 °C heat-treatment, but is absent after the 1100 °C one, giving us thresholds for the precipitation and dissolution temperatures of this phase.

For Cu-containing alloys, a secondary FCC phase with different lattice parameters is observable after the 600 °C treatment for the CuCoCrFeNi alloy and after the 800 °C for the AlCuCoCrFeNi.

All samples exhibit an increase in hardness after the first 600 °C treatment, even AlCuCoCrFeNi, within which the concentration of the soft FCC phase increases. This can be explained by a better dissolution of the various elements in the alloy's lattice, increasing solid solution strengthening [12]. Except for $Al_{0.5}$CoCrFeNi, which hits a hardness maximum after the 800 °C treatment due to the precipitation of the BCC and sigma phase, all alloys' hardness decreases for treatment temperatures beyond 600 °C. These subsequent decrease in the measured microhardness could be explained by grain coarsening [13], but this will have to be verified by further microstructural analysis.

4 Discussion

4.1 Impact of Al and Cu Concentration on the Mechanical Properties

The measured hardness increase with Al concentration was observed before for the AlxCoCrFeNi family obtained through vacuum arc melting and casting, as well as by Laser Metal Deposition method [14]. This behaviour remains visible when an LPBF process is used. Here, this hardness increase can be explained by the transition of the alloy from an FCC phase to a BCC phase, which was already observed for the conventionally alloyed AlxCoCrFeNi family [14]. It can also be explained by the fine granular structure observed for the equimolar AlCoCrFeNi alloy. However, the process induced cracks and pores led to very brittle behaviours and poor tensile properties for AlCoCrFeNi, AlCuCoCrFeNi and CuCoCrFeNi. For the latter, the high crack concentration can be explained by the precipitation of a brittle intergranular Cu-rich phase [15].

AlCoCrFeNi and AlCuCoCrFeNi present similar properties and microstructure. Likewise, there are no significant differences between the properties of CuCoCrFeNi and the properties of the reference Cantor alloy. This suggests that the Al concentration has a much higher impact on the properties of the CoCrFeMnNiAlCu family than the Cu concentration does.

As expected, the hardness obtained for the Cantor alloy is similar to what can be observed for an LPBF 316L stainless steel. Al0.3CoCrFeNi, Al0.5CoCrFeNi and CuCoCrFeNi also have hardness of similar magnitude. However, for the BCC-phase HEA (AlCoCrFeNi and AlCuCoCrFeNi), their hardness is even higher than what can be achieved with an LPBF SAF 2507 Super Duplex alloy, which is also fully BCC after LPBF manufacturing.

4.2 LPBF Induced Phase Repartition

The fast cooling rate inherent to laser fusion processes was shown to hamper the formation of secondary phases in otherwise dual-phase alloys. This can diminish the overall attractivity of these alloys, as their main advantage is to have a good strength/ductility trade-off thanks to their dual FCC/BCC structure. This issue could be solved with one of the thermal treatments applied to the samples, and remains to be verified through proper tensile tests.

5 Discussion

This study analyses the effect of a progressive variations of the Cantor alloy's composition, using multiple different concentrations of Al and/or Cu, on their LPBF printabilities, microstructures and mechanical properties. It led to the following conclusions:

1/ The observed phases in the as-printed state were FCC, BCC/B2 or a combination of both. For the later, only one of the two phases were visible on the EBSD phase maps, this being supposedly due to the extreme cooling speeds inherent to the LPBF process.

2/ Adding Al promotes the formation of a BCC/B2 phase, while also improving the hardness and the tensile properties up until $Al_{0.5}$. Going beyond this Al concentration leads to a significant degradation in tensile properties due to cracks and pores. Adding Copper does not change the FCC phase observed in the reference Cantor alloy, nor significantly changes the mechanical properties.

3/ Heat treatment leads to the formation of a σ-phase at 800 °C for Al-containing HEAs, which disappears after a 1100 °C treatment. For Cu-containing HEAs, a secondary FCC phase with different lattice parameters appears. Overall, hardness increases after a 600 °C treatment, which can be explained by a better dissolution of the elements in the matrix leading to the increase of solid solution strengthening phenomena.

Acknowledgements. This work is supported by the French Defense Innovation agency (AID) from the French Army Ministry. The authors also gratefully acknowledge the helpful comments and suggestions of the reviewers, which have improved the presentation.

References

1. Zhang, Y., et al.: Microstructures and properties of high-entropy alloys. Prog. Mater. Sci. **61**, 1–93 (2014). https://doi.org/10.1016/j.pmatsci.2013.10.001
2. Cantor, B., Chang, I.T.H., Knight, P., Vincent, A.J.B.: Microstructural development in equiatomic multicomponent alloys. Mater. Sci. Eng. A **375–377**, 213–218 (2004). https://doi.org/10.1016/j.msea.2003.10.257
3. Zhang, Y., Yang, X., Liaw, P.K.: Alloy design and properties optimization of high-entropy alloys. JOM **64**, 830–838 (2012). https://doi.org/10.1007/s11837-012-0366-5
4. Ye, Y.F., Wang, Q., Lu, J., Liu, C.T., Yang, Y.: High-entropy alloy: challenges and prospects. Mat. Today **19**, 349–362 (2016). https://doi.org/10.1016/j.mattod.2015.11.026
5. Shams, S.A.A., et al.: Low-cycle fatigue properties of CoCrFeMnNi high-entropy alloy compared with its conventional counterparts. Mater. Sci. Eng. A **792**, 139661 (2020). https://doi.org/10.1016/j.msea.2020.139661
6. Bartolomeu, F., et al.: 316L stainless steel mechanical and tribological behavior—A comparison between selective laser melting, hot pressing and conventional casting. Addit. Manuf.. Manuf. **16**, 81–89 (2017). https://doi.org/10.1016/j.addma.2017.05.007
7. Xian, X., et al.: Precipitation and its strengthening of Cu-rich phase in CrMnFeCoNiCux high-entropy alloys. Mat. Sci. Eng. A **713**, 134–140 (2018). https://doi.org/10.1016/j.msea.2017.12.060
8. Zhu, Z.G., et al.: Hierarchical microstructure and strengthening mechanisms of a CoCrFeNiMn high entropy alloy additively manufactured by selective laser melting. Scripta Mater. **154**, 20–24 (2018). https://doi.org/10.1016/j.scriptamat.2018.05.015
9. Zhou, P.F., Xiao, D.H., Wu, Z., Ou, X.Q.: Al0.5FeCoCrNi high entropy alloy prepared by selective laser melting with gas-atomized pre-alloy powders. Mater. Sci. Eng. A **739**, 86–89 (2019). https://doi.org/10.1016/j.msea.2018.10.035
10. Karlsson, D., et al.: Elemental segregation in an AlCoCrFeNi high-entropy alloy – a comparison between selective laser melting and induction melting. J. Alloy. Compd. **784**, 195–203 (2019). https://doi.org/10.1016/j.jallcom.2018.12.267
11. Gao, J., et al.: Fabricating antibacterial CoCrCuFeNi high-entropy alloy via selective laser melting and in-situ alloying. J. Mater. Sci. Technol. **102**, 159–165 (2022). https://doi.org/10.1016/j.jmst.2021.07.002
12. Lee, C., et al.: Lattice distortion in a strong and ductile refractory high-entropy alloy. Acta Mater. **160**, 158–172 (2018). https://doi.org/10.1016/j.actamat.2018.08.053

13. Callister, Jr.: WD Fundamentals of Materials Science and Engineering, 5th edn. John Wiley & Sons, Inc.
14. Wang, W.-R., Wang, W.-L., Wang, S.-C., Tsai, Y.-C., Lai, C.-H., Yeh, J.-W.: Effects of Al addition on the microstructure and mechanical property of AlxCoCrFeNi high-entropy alloys. Interm. **26**, 44–51 (2012). https://doi.org/10.1016/j.intermet.2012.03.005
15. Du, C., Hu, L., Pan, Q., Chen, K., Zhou, P., Wang, G.: Effect of Cu on the strengthening and embrittling of an FeCoNiCr-xCu HEA. Mater. Sci. Eng. AA **832**, 142413 (2022). https://doi.org/10.1016/j.msea.2021.142413

Mechanical Properties of Additively Manufactured 17-4PH SS: Heat Treatment

Thabet A. M. Sghaier[1](✉), Habib Sahlaoui[1], Tarek Mabrouki[2], Haifa Sallem[3], and Joël Rech[4]

[1] Laboratory of Mechanics, Materials and Processes (LMMP), National High School of Engineering of Tunis (ENSIT) , University of Tunis (UT), LR 99ES05, 5 Avenue Taha Hussien, B.P 56, 1008 Bab Menara, Tunisia
thabetsgaier@gmail.com, habib.sahlaoui@ensit.rnu.tn

[2] Applied Mechanics and Engineering Laboratory, National Engineering School of Tunis (ENIT), University of Tunis El Manar, (LMAI-LR-11ES19), Le Belvedere, BP-37, 1002 Tunis, Tunisia
tarek.mabrouki@enit.utm.tn

[3] Powder Technology and Advanced Materials, HEI-VS, HES-SO University of Applied Sciences and Arts Western Switzerland, Rue de l'Industrie 23, 1950 Sion, Switzerland
haifa.sallem@hevs.ch

[4] Laboratoire de Tribologie et Dynamique des Système (LTDS-UMR5513), ECL-ENISE, 58, rue de Jean Parot, 42023 Saint-Etienne Cedex 2, France
joel.rech@enise.fr

Abstract. The effects of the thermal post-treatments on the mechanical properties of SLMed 17-4PH stainless steel are studied. Three thermal post-treatments are carried out: (E0) as received state, (E1 and E2) solution annealing treatment and (E3) aging treatment. Microhardness, impact-strength, and wear tests are carried out to determine the effect of treatment on the mechanical properties of SLMed 17-4PH. The results showed that the hardness of 17-4PH at the E0 state is a greater than the hardness of the powder. This hardness decreases slightly from 418 HV to 384 HV after the solution annealing treatment E2 and increase again after E3 ageing treatment to around 447HV. A hardness difference between the two manufacturing directions was also observed. This difference almost disappears after the E3 ageing treatment. The impact strength results show a huge drop in impact strength for 17-4PH obtained by SLM. An improvement in wear behaviour following heat treatment, especially in the case of ageing.

Keywords: 17-4PH Stainless Steel · Selective Laser Melting · heat post-treatment · hardness · impact strength

1 Introduction

Additive Manufacturing (AM) by Selective Laser Melting (SLM) has reached most industrial sectors thanks to its ability to manufacture parts and structures with more or less complex geometries, its high precision and its cost reduction and materials savings.

T. Mabrouki et al. (Eds.): AIAM 2023, LNME, pp. 28–36, 2024.
https://doi.org/10.1007/978-3-031-47784-3_4

This explains the great interest given to the integration of SLM process in the development of stainless steel parts in order to benefit from the multiple advantages offered by this process: geometric flexibility, rapid production, control of the microstructure via the optimization of the process parameters, etc. Nevertheless, SLM sometimes has undesirable effects on the mechanical and physicochemical properties and consequently on the service behavior of stainless steel parts. Indeed, during SLM manufacturing, the selective melting process is complex and rapid, in which many intrinsic and extrinsic factors interfere, causing several mechanical and microstructural defects that reduce the quality of the manufactured parts. In addition, the chemical heterogeneities of the raw material associate at the severe temperature gradients which is usually followed followed by high cooling speed generate exceptionally heterogeneous microstructures, residual stresses, poor roughness, anisotropy, etc., This highlights the prospects of AM specifics steels development taking advantage of the typical thermal cycles associated with the SLM process. Thus, the use of SLM to manufacture steel parts and structures are promising, but more work is needed so that it can replace conventional processes (Sghaier, T.A.M. 2023). In particular, the understanding of the unique microstructural evolution during manufacturing by SLM is essential to be monitoring the parameters process according to the mechanical and physicochemical properties required and to develop new thermal post-treatment programs which could be different from conventional ones.

Several research works have been carried out during this last decade on the post-treatments of several metals obtained by AM. Among other, the effects of proposed thermal (Lashgari et al. 2023), thermochemical (Jessy Michla et al. 2022) and mechanical (AlMangour and Yang 2016) post-treatments on certain properties of 17-4PH stainless steel obtained by SLM have been studied.

2 Experiments

2.1 SLM process

The 17-4PH stainless steel powder used in this study is of an average of 60 μm in diameter. The chemical composition of the powder in % of weight is listed in Table 1 (ASTM A364).

Table 1. Chemical composition of the 17-4PH powder (wt%).

Composition (wt%)	C	Si	Mn	Ni	Cr	Cu	Co	Nb + Ta
17-4PH	< 0.07	< 1	< 1	3–5	15–17.5	3–5	0.03	0.15–0.45

The selective laser melting of the specimens is carried out on a PHENIX SYSTEMS type PROX200 machine. The specimens are printed in a chamber measuring 140×140× 125 on a 304L stainless steel support plate. The manufacturing chamber is maintained under a protected atmosphere (Argon (Ar)). All samples were processed with a 10 mm hexagonal lasering strategy in parallel lines forming crosshatches between each fused layer, a rotation with an angle of 90° was performed (odd layer 0°, even layer 90°) with an overlap of the hexagons of 100 μm. The used laser scanning strategy during SLM manufacturing is showed by Fig. 1. Two types of specimens were produced in the same tray: prismatic with dimensions of 12 × 12 × 57 mm^3 and cylindrical with dimension Ø12 mm × 65 mm. The used processing parameters are given in Table 2.

After the parts have been manufactured and before they are disassembled, they subjected to a stress relief annealing, i.e., 760 degrees for 2 h followed by slow cooling in air (State E0). The annealing treatment lead to minimize residual stresses in the structure and reduce the risk of dimensional changes during subsequent manufacturing operations. The specimens are finally cut out of the manufacturing plate by means of an electro discharge machining (EDM).

Table 2. SLM processing parameters.

$E = 64\ Jmm^{-3}$	
SLM parameters	Value
Power (w)	240
Laser Scan speed (mm/s)	2500
Hatching distance (μm)	50
Layer thickness (μm)	30

2.2 Heat Treatments

The normalized samples are heat treated in a resistance furnace (Protherm Furnace model FLP 120/10). In order to study the influence of thermal post-treatments on the mechanical properties of 17-4PH stainless steel manufactured by SLM, three thermal treatments were applied:

- **E0:** SLM as received: stress relief annealing at 760 degrees for 2 h followed by slow cooling in air
- **E1:** Solution annealing heat treatment at 1050 °C for 1 h finished with water quenching.
- **E2:** Solution annealing heat treatment at 1050 °C for 1 h followed by air cooling.
- **E3:** Solution annealing heat treatment at 1050 °C for 1 h followed by air cooling and aging treatment (H1150) tempering at 620 °C for 4 h followed by air cooling.

These treatments were applied to the SLM samples according to the ASTM A564 standard recommended for structurally hardened steel.

2.3 Mechanical Characterization

The mechanical properties of these samples are assessed by Charpy impact, micro-hardness and micro-scratch tests.

- FRANK (300J) pendulum (Fig. 1) was used to evaluate the conventional impact strength energy. The investigated specimens are prepared according to NF EN ISO 148–1 (2017) with a 55 mm of lengths, 10×10 mm^2 of nominal cross-sections, 2 mm depths of lateral V-notch and 0.25 mm of curvature radii.

Fig.1. FRANK (300J) pendulum used in this study.

- The hardness of the samples was measured using a Vickers microhardness tester FM-810 (Fig. 2) with a load of 50 g. A minimum of 10 measurements were performed at on the polished surfaces.

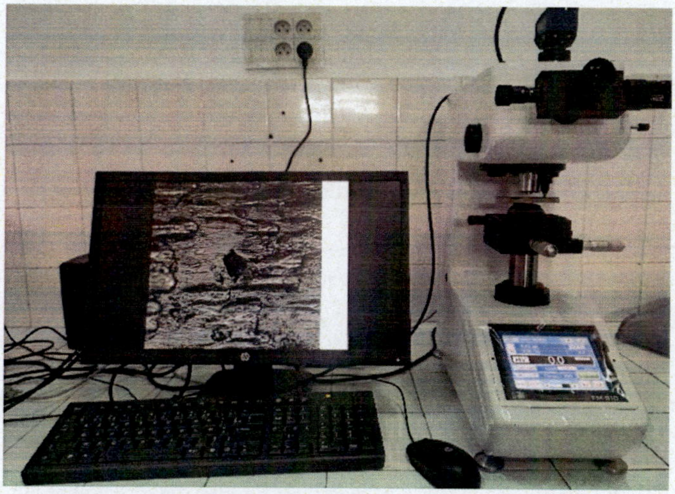

Fig. 2. Microhardness tester FM-810 used in this study.

- CSM instruments (Fig. 3) were used to perform dry sliding micro-scratch tests using a diamond Rockwell-type indentation with a radius of 200 μm. After polishing the surfaces, 20-pass linear scratch tests were performed with applied test loads of 1 N, 3 N and 5 N.
- A minimum of twenty straight lines were drawn per sample, the coefficient of friction (COF) versus time was recorded and the cross-sectional area of the wear track below reference line was used to evaluate the wear volume (V_f in mm^3).

Fig. 3. CSM instruments used for micro-scratch tests.

3 Results and Discussion

3.1 Microhardness and Impact-Strength Energy Measurements

The variation of the Vickers microhardness of the SLM as-received, and thermally treated states is plotted in Fig. 4.

The micro-hardness of the E0 state is of the order of 439 ± 27 HV0.5 and it varies according to the direction of construction. Indeed, the mean value of 460 ± 20.5 HV0.5 obtained in the plane is slightly higher than that measured along normal (construction direction) which is of 418 ± 29.5 HV0.5. This difference can be attributed to microstructure anisotropy in both directions (AlMangour and Yang 2016, Cheng et al. 2020). A significant decrease in average microhardness after the solution treatment and cooling with air and water (E1 and E2) respectively 396 ± 21.5 HV0.5; 412 ± 25.5 HV0.5. In addition, the difference in hardness between XZ plane (384 ± 23.5 HV0.5) and XY plane (409 ± 25 HV0.5) of the sample treated by E1 also became lower while in E2 condition the difference increased a little compared to SLM due to the post-heating treatments.

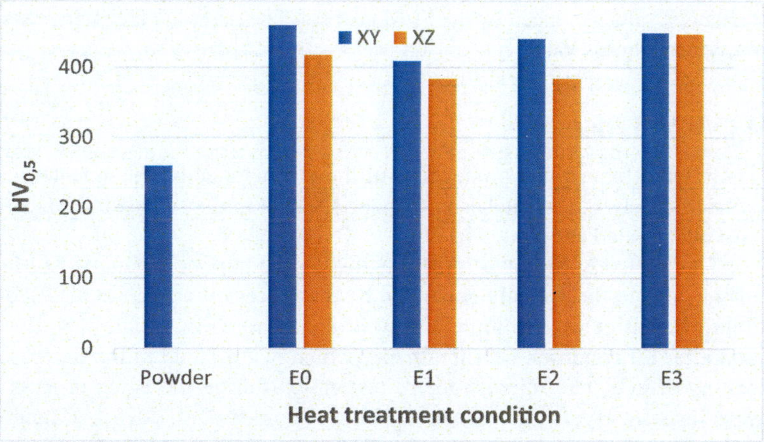

Fig. 4. Microhardness of 17-4PH powder, SLMed as-received and heat-treated state along the horizontal (XY plane) and vertical (XZ plane) directions.

The average microhardness of the solution aged sample (E3) is higher than all other sample states, moreover the difference in micro hardness between the two manufacturing planes is almost attenuated.

The hardness profile along the construction axis (Z) shows that the hardness slightly increases (Fig. 5), this can be explained by the difference of thermal gradient between the construction plate and the part to be manufactured at the beginning, then a lack of fusion accompanies the beginning of the SLM process, moreover, there is an accumulation of temperature which serves to fuse the totality of powder thus increasing the density of the part.

Unexpected results appear nevertheless: an enormous drop in impact toughness for SLM as built (~11 J/cm2). The post-heat treatment cannot significantly repair the low

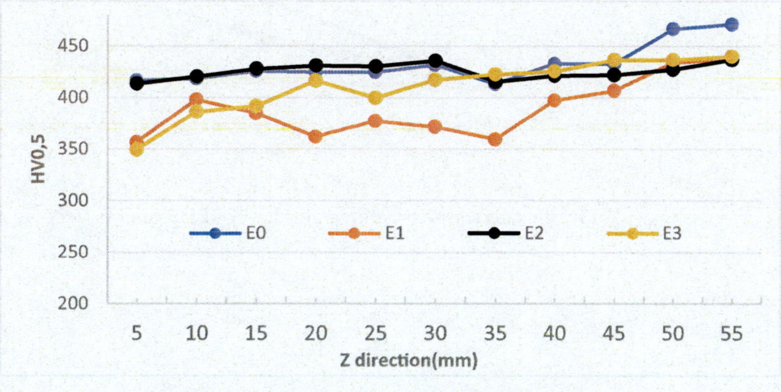

Fig. 5. Hardness profile along the building axis (z).

resilience values induced by the SLM process. The pores are the main reason for the low impact toughness value.

3.2 Wear Performance

The wear volume variation depends of the load applied by the indenter and coefficient of frictions after 20 scratches and under 1N load in the samples SLM, as-received and post heat treated are plotted in Fig. 6.

The relatively high wear volume loss in the surface layers of SLMed specimens may be due to the porosity generated by the process (Lashgari et al. 2020). Significant improvement in wear volume and wear behaviour (Fig. 6 and 7) in the SLM condition after E3 condition heat treatment, which is attributed to the decrease of porosity, balling defects, and microstructure restoration during the aging process that forms a very resistant layers to wear and spalling (Kempen et al. 2015; AlMangour and Yang 2016).Therefore, the aging treatment E3 leads to the formation of a tribolayer hardened and to the improvement of the wear performance of the part.

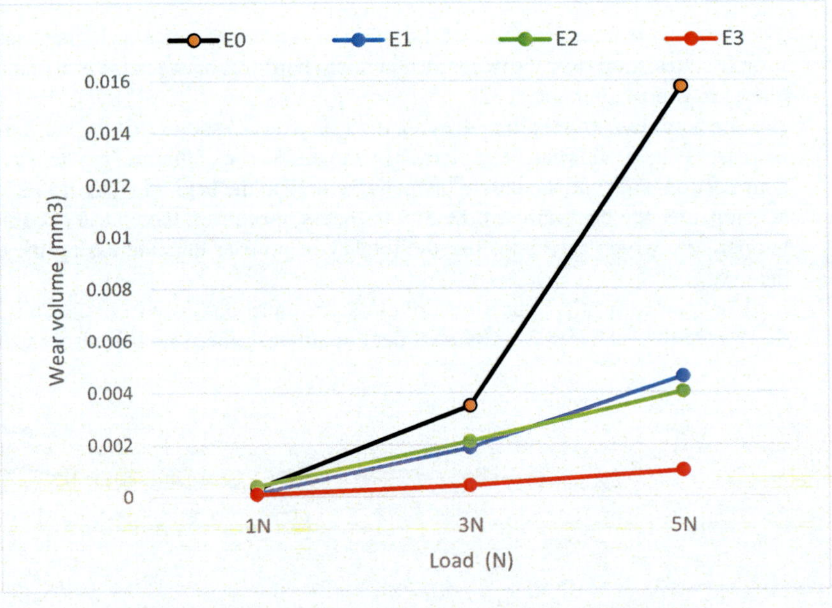

Fig. 6. Average wear volume for the 17-4PH stainless steel

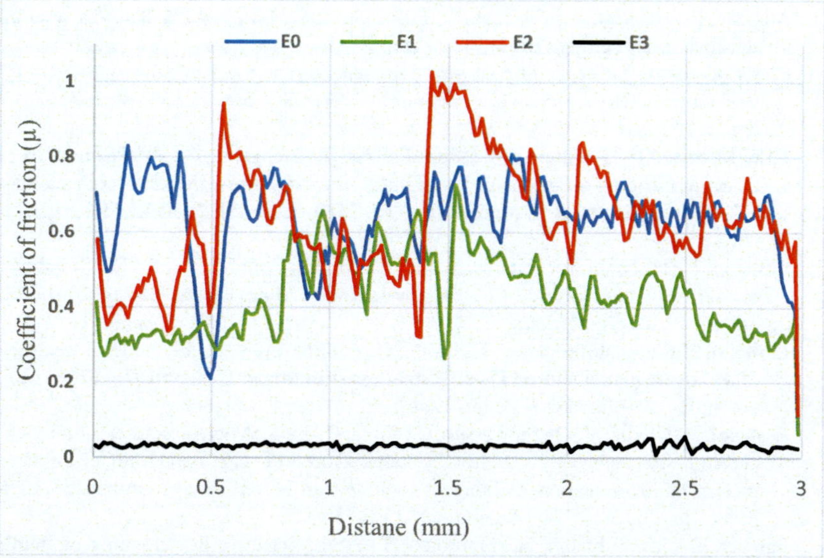

Fig. 7. The variation of coefficient of friction along the build direction to the build direction (20 scratch, 1N)

4 Conclusion

In this work, the effect of the post-heat treatments on the hardness, the impact strength and the wear of additively manufactured 17-4PH steel obtained by SLM is studied in four states (E0, E1, E2 and E3) and the most following conclusions can be drawn:

1- The 17-4PH produced by SLM has a much higher hardness than the powder. The post-treatments applied led to a reduction in hardness and a reduction in the variation in micro-hardness between the two manufacturing directions (XY and XZ) after ageing treatment.

2– 17-4PH obtained by SLM has a much lower impact strength than that obtained by conventional processes. Heat treatments cannot compensate for this lower impact strength.

3- Significant variation of the coefficient of friction along the construction direction was shown. It is probably attributed to the poor surface roughness of the parts obtained by SLM.

4- Heat treatments can significantly improve wear behaviour, and aged 17-4PH has better wear behaviour than other treatments.

References

AlMangour, B., Yang, J.-M.: Improving the surface quality and mechanical properties by shot-peening of 17–4 stainless steel fabricated by additive manufacturing. Mater. Des. **110**, 914–924 (2016). https://doi.org/10.1016/j.matdes.2016.08.037

Auguste, P., Mauduit, A., Fouquet, L., Pillot, S.: Study on 17–4 PH stainless steel produced by selective laser melting (2018)

Chang, C., et al.: Influence of post-heat treatments on the mechanical properties of CX stainless steel fabricated by selective laser melting. J. Mater. Sci. **55**(19), 8303–8316 (2020). https://doi.org/10.1007/s10853-020-04566-x

Jessy Michla, J.R., Ravikumar, B., Ram Prabhu, T., Siengchin, S., Arul Kumar, M., Rajini, N.: Effect of nitriding on mechanical and microstructural properties of direct metal laser sintered 17–4PH stainless steel. J. Market. Res. **19**, 2810–2821 (2022). https://doi.org/10.1016/j.jmrt.2022.05.198

Lashgari, H.R., Adabifiroozjaei, E., Kong, C., Molina-Luna, L., Li, S.: Heat treatment response of additively manufactured 17–4PH stainless steel. Mater Charact **197**, 112661 (2023). https://doi.org/10.1016/j.matchar.2023.112661

Lashgari, H.R., Adabifiroozjaei, E., Kong, C., Molina-Luna, L., Li, S.: Heat treatment response of additively manufactured 17–4PH stainless steel. Mater Charact **197**, 112661 (2023). https://doi.org/10.1016/j.matchar.2023.112661

Lashgari, H.R., Xue, Y., Onggowarsito, C., Kong, C., Li, S.: Microstructure, tribological properties and corrosion behaviour of additively manufactured 17–4PH stainless steel: effects of scanning pattern, build orientation, and single vs double scan. Mater. Today Commun. **25**, 101535 (2020). https://doi.org/10.1016/j.mtcomm.2020.101535

Sghaier, T.A.M., Sahlaoui, H., Mabrouki, T., et al.: Selective laser melting of stainless-steel: a review of process, microstructure, mechanical properties and post-processing treatments. Int. J. Mater. Form. **16**, 41 (2023). https://doi.org/10.1007/s12289-023-01769-w

Nitinol Stents Printed by Selective Laser Melting

Jean Pralong[1], Livia Jerjen[1], Bruno Schnyder[1], Oksana Banakh[2], Tony Journot[2], Haifa Sallem[1], and Samuel Rey-Mermet[1](✉)

[1] HEI-VS, University of Applied Sciences Western Switzerland HES-SO, Sion, Switzerland
Samuel.rey-mermet@hevs.ch

[2] HE-Arc Ingénierie, University of Applied Sciences Western Switzerland HES-SO, La Chaux-de-Fonds, Switzerland

Abstract. Nitinol alloys stents have been printed by Selective Laser Melting from a gas atomized powder with a composition of 50.8 at% Ni. The energy density has been adapted to set the austenite finish temperature just below the human body temperature, therefore the stent diameter can be reduced by plastic strain at lower temperature and recover its former expanded shape after being inserted in the artery. Several mesh geometries have been mechanically analyzed by finite elements modeling to ensure that the local strain will not exceed superelastic reversible capacity of 4% while squeezing the stent diameter by a factor of 4. The stent surface has been treated by electropolishing to reduce its roughness to Ra < 0.02 mm. After polishing, some samples have been coated by a 50 nm thick TiO_2 layer by Atomic Layer Deposition. Biocompatibility and hemocompatibility analysis demonstrated that the TiO2 coating significantly improved hemocompatibility. Nevertheless, uncoated stents are also biocompatible and hemocompatible according to ISO-10993–12 guidelines.

Keywords: Selective laser melting · nitinol · biocompatibility · superelastic behavior · shape memory behavior

1 Introduction

Nitinol alloys (Ni50-Ti50) have found their niche in the medical industry since the 1990s. Cardiovascular or cerebrovascular implants (stents, catheters) represent more than 50% of this market and wire guides about 10%. Recent applications include a growing interest in the orthopaedic sector for Nitinol in the manufacture of staples, bone screws and other implants. The alloy is used for its specific characteristics of shape memory and superelasticity (10 times higher than traditional metals). The transition temperature between these two properties must be adapted to the desired function. In addition, nitinol has good corrosion resistance, a Young's modulus close to that of bone and is biocompatible (Parvizi et al. 2021).

Stent fabrication process includes several deformation steps coupled with thermal treatment to obtain tube or wire. Then the wires are welded together, or the tubes are laser cut to obtain a cylindrical mesh called stent. Nitinol properties are very sensitive to

T. Mabrouki et al. (Eds.): AIAM 2023, LNME, pp. 37–43, 2024.
https://doi.org/10.1007/978-3-031-47784-3_5

thermal history and once shaped, the stent undergoes a specific thermal treatment to reach the required properties. The most critical of these, is the transition temperature between the austenitic and martensitc phases. This defines the temperature under which the alloy exhibits superelasticity and above which it will recover its former shape if plastically strained. This phase transition is also very sensitive to the nickel atomic content. A strict control of the alloy composition during selective laser melting (SLM) production, as well as during laser machining, is therefore imperative (Bayati et al., 2020).

Apart from the nitinol composition control during the SLM process, the main remaining objective is that the printed parts can do not impinge human health safety. This requires:

- a minimal surface roughness,
- that the stent can be plastically strained before the insertion in the artery and recovers its expanded shape below 37°C,
- that the stent is biocompatible and hemocompatible
- and that all these properties remain stable for 10^7 cycles under a 1% elastic strain amplitude.

2 Methods

Stents have been printed by SLM with an SLM Solutions 125 printer by using an atomized nitinol powder with a Dv50 of 35 μm and a composition of 50.8% at. of nickel. Printing parameters such as power, hatching space and scan velocity have been varied to obtain volumetric energy densities ranging between 50 and 250 mJ/mm3. Base plate temperature was of 200 °C and the part was printed under a protecting argon flow.

A plastic radial deformation of the stent is required before being inserted into the artery. The maximal local strain of the mesh must remain below the shape recovery strain of 4% measured on printed nitinol tensile probes. Thus, several mesh patterns have been tested by applying a radial displacement to a stent corresponding to a diameter reduction of a factor of 4. These calculations were performed by using ANSYS.

Nitinol samples of approximately 30 μg have been cut from printed stents and cubes to measure the transition temperatures by differential scanning calorimetry (DSC) with Mettler Toledo DSC823e calorimeter. Thermal cycles between -100 °C and + 100 °C, during heating, cooling and second heating at a rate of ± 10 °C/minute with 1' plateau at -100 °C and + 100 °C have been performed.

Cubes and stents densities have been measured by the Archimedes method in water.

The stent surface has been treated by electropolishing (EP) to reduce its roughness. Two stainless steel 3D printed meshes, having the same patterns as the nitinol stents have been used as outer and inner counter-electrodes as seen on the Fig. 1. A parametric study has been performed to establish EP conditions: electrolyte molar ratios of KCl:ethylene glycol between 1 to 5%, temperature of 80 °C or 130 °C with an applied voltage of 60V.

After polishing, some stents have been coated by a 50 nm thick TiO_2 layer by Atomic Layer Deposition.

Cytotoxicity in L929 cell cultures and hemocompatibility by osmotic fragility assay have been performed on as-printed, polished, and coated stents according to ISO-10993–12 guidelines.

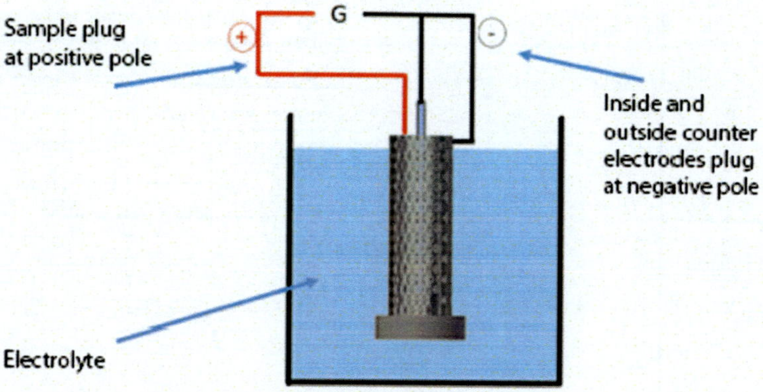

Fig. 1. Electropolishing setup

3 Results and Discussion

Our project demonstrates the potential of SLM technology for the fabrication of nitinol implants with novel functionalities. For this purpose, the SLM operating conditions will be optimized to produce superelastic or shape memory implants and to guarantee the alloy composition by controlling nickel evaporation during printing. An electropolishing treatment will be applied to improve the surface quality. It is necessary to remove weakly attached powder particles from the surface of the parts. After this first post treatment, a thin layer of TiO_2 will be deposited by atomic layer deposition (ALD) to improve biocompatibility and will have to resist superelastic deformation (Speirs et al., 2017).

Hemocompatibility and biocompatibility of the coated, polished, and uncoated parts has been tested. The results show that all 3 tested samples are biocompatible and hemocompatible.

As previously shown in other studies such as (Bayati et al., 2021), the transition temperature (TT) between austenitic and martensitic phases are related to the printed parameters. TT variations have already been observed during selective laser melting of nitinol and have been related to the evaporation of the nickel from the melt pool, as it is well known that richer Ti alloys show higher TT (Patoor & Berveillier, 1994).

As shown on Fig. 2, the austenite finish Af and martensite finish Mf temperatures are slightly increasing with the volumetric energy density Ev. But for identical Ev values, there is a significant drop of TT when the laser power is increased, especially from 100 W to 250 W as seen one the Fig. 2. This seems to indicate that Ni evaporation is mostly related to the melt pool characteristics such as its size and temperature, and that evaporation is accelerated when these values reach a given threshold as previously described in (Moghaddam et al., 2019).

Stents and cubes ($4 \times 4 \times 6$ mm) printed with identical parameters show almost identical TT, as shown on Fig. 3 for A_f. This seems to confirm that the part size, its influence on the thermal gradient and cooling time only have a slight influence on TT. The relation between the melt pool characteristics and the TT requires further investigations to be fully understood.

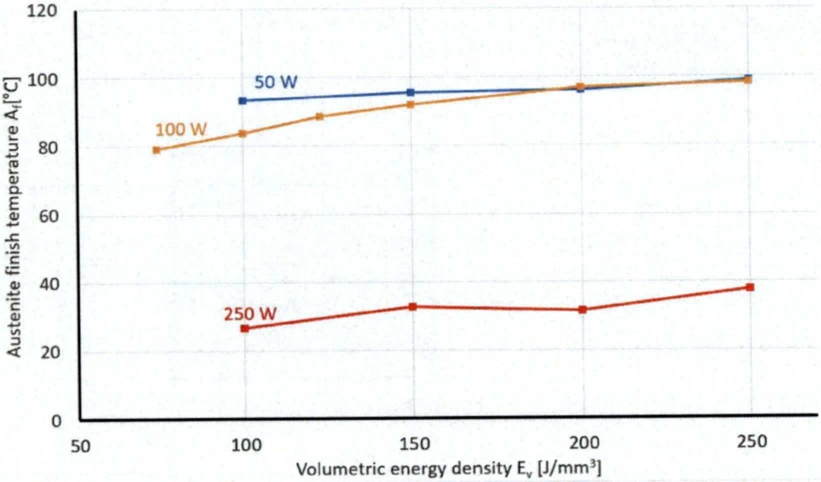

Fig. 2. A_f measured by DSC on stents printed with different volumetric energy densities (100 to 250 W/mm3) and laser powers (50, 100 or 250 W).

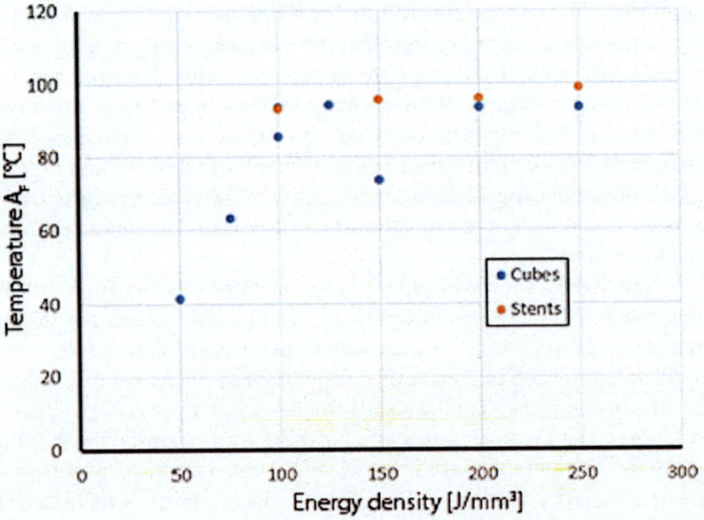

Fig. 3. A_f measured by DSC on stents and cubes printed with different energy densities.

In this work, parameters to adapt A_f between 27 and 93 °C have been identified and stents with A_f slightly below 37 °C have been build. All these samples show relative densities greater than 98%.

Printed stents have a minimal mesh diameter of 200 or 300 μm and internal diameter of 6 or 9 mm. The mesh diameter is limited by the SLM process resolution. The inner diameter has been chosen to match the EP counter electrode dimensions but can be reduced further. Several mesh patterns are shown on Fig. 4.

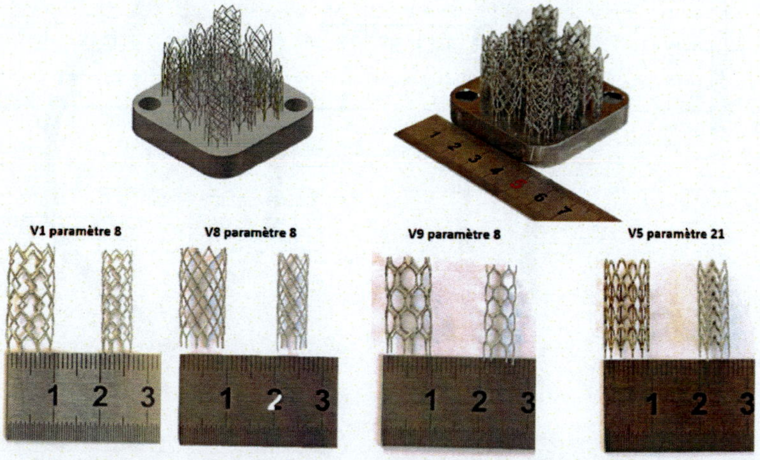

Fig. 4. Examples of printed nitinol stents geometries

The Fig. 5 shows a mesh pattern that limits the peak strain to 4% when the stent diameter is squeezed by a factor 4, corresponding to the contraction required to insert the sent in an artery.

Electropolishing helps to reduce surface roughness to 13 nm (ISO 21920) on flat samples. For 3D structures polishing, stainless steel inner and outer counter electrodes have been printed by SLM. It has permitted to reduce the roughness significantly on the outside and the inside of the stents as shown on Fig. 6. Due to the curvature of these surfaces, it has not been possible to measure the roughness precisely.

The results show that TiO_2 coated, uncoated and polished stent extracts do not have an impact on the growth of the tissue cells (L929 fibroblasts). Thus, they show no cytotoxic effect, indicating promising tissue biocompatibility. TiO2-coated and uncoated samples seem to have a minor impact on red blood cells, inducing 10–20% hemolysis. However, this is acceptable, as clinical reference values suggest that a reduction of up to 26.5% is tolerable in humans. The normal levels of hemoglobin are 133-177g/L for men, and 117-157g/L for women. Anemia is defined as a hemoglobin level < 130g/L and < 115g/L respectively, thus a reduction of max. 26.5% for both in hemoglobin and consequently erythrocytes can be tolerated by the body without becoming anemic (Schmidt et al., 2014).

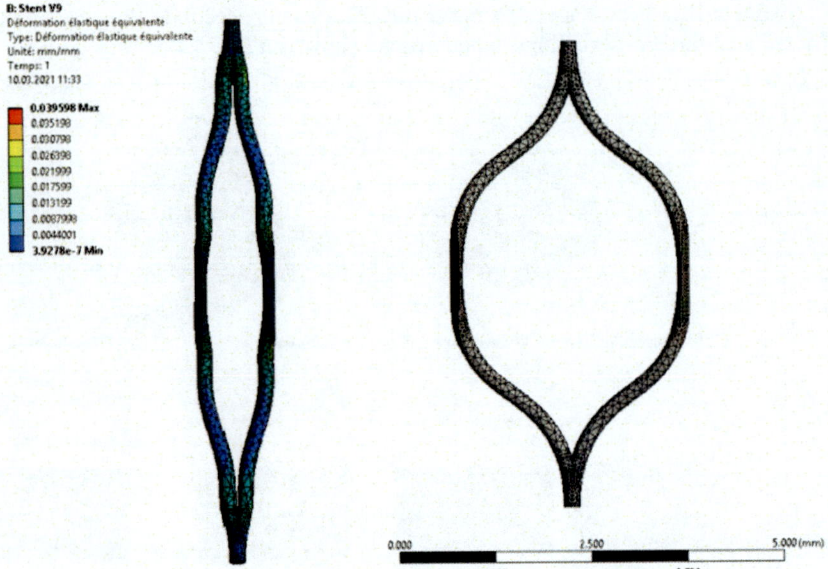

Fig. 5. FEM calculated strain in a single mesh for a stent diameter contraction of a factor 4.

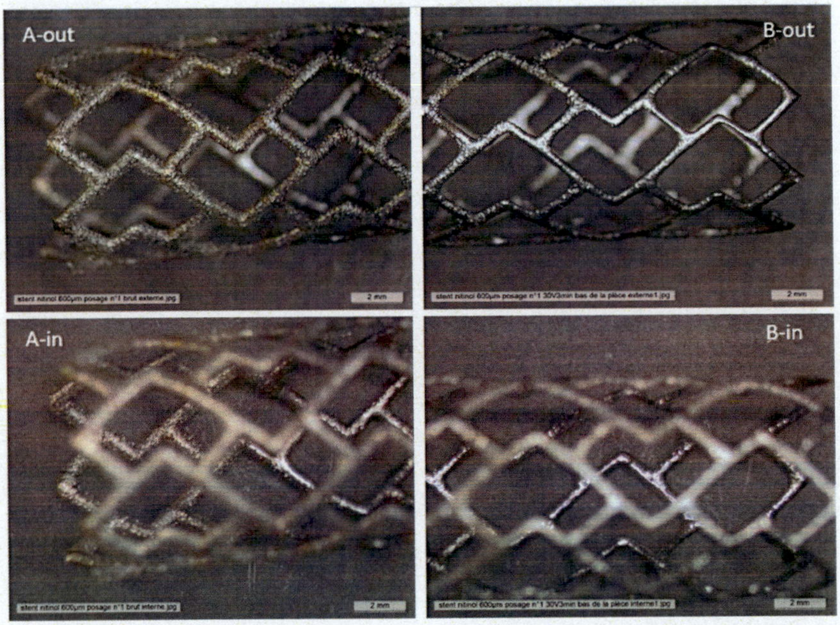

Fig. 6. A-out and A-in: outside and inside view of the mesh roughness as printed. B-out and B-in same views after EP

4 Conclusion

Nitinol stents have been successfully printed by SLM from a Ni rich powder (50.8%at. Ni). The alloy composition has been adapted by changing the process parameters. The austenite finish temperature has been varied between 27 and 93°C, and thus stents showing a full shape memory recovery at body temperature can be produced. Finite element modelling calculations have permit to design a specific mesh pattern limiting the maximal strain to 4% while the stent diameter is reduced by a factor of 4. This value corresponds to maximal recoverable strain measured on SLM dog bones sample printed in the same conditions. The roughness has been reduced by electropolishing on the outer and inner diameter of the stent by using two specifically designed counter electrodes. The deposition of a 50 nm thin TiO_2 layer significantly decreases the red blood cell mortality, but uncoated and even unpolished stents also exhibit hemocompatibility and biocompatibility.

The printed stents have demonstrated shape memory effect and superelasticity. These properties have not be precisely measured and the stent have not been tested under cyclic strains.

Nevertheless, the mesh diameter is limited by the process resolution to 200 μm. This will remain a major drawback for an industrial production. A possible solution to be considered is the use of electropolishing to reduce the stent mesh diameter. Another drawback is the requirement of specific counter electrodes during this step. This will limit the shape flexibility of the SLM method. Nevertheless, the possibility to print composition graded nitinol devices with mesh geometries is of a great interest for specific domains such as the space industry.

Acknowledgements. The authors want to acknowledge the University of Applied Sciences and Arts of Western Switzerland (HES-SO) for funding this research.

References

Parvizi, S.P., Mahdi, H.S., Asgarinia, F., Nematollahi, M., Elahinia, M.: Effective parameters on the final properties of NiTi-based alloys manufactured by powder metallurgy methods: a review. Prog. Mater. Sci. **117**, 100739 (2021)

Bayati, P., et al.: Toward low and high cycle fatigue behavior of SLM-fabricated NiTi: considering the effect of build orientation and employing a self-heating approach. Int. J. Mech. Sci. **185** 105878 (2020)

Speirs, M., Van Hooreweder, B., Van Humbeeck, J., Kruth, J.-P.: Fatigue behaviour of NiTi shape memory alloy scaffolds produced by SLM, a unit cell design comparison. J. Mech. Behav. Biomed. Mater. **70**, 53–59 (2017)

Moghaddam, N.S., et al.: Achieving super elasticity in additively manufactured NiTi in compression without post-process heat treatment. Nat. Sci. Rep. **9**, 14 (2019)

Patoor, E., Berveillier, M.: Technologie des alliages à mémoire de forme, Ed. HERMES, Paris (1994)

Schmidt, P., Cornu, P., Angelillo-Scherrer, A.: Bases physiopathologiques en hématologie générale. Un aide-mémoire d'hématologie. Service d'Hématologie, CHUV, Lausanne (2014)

Preliminary Study of Geometric Defects and Topography of Bio-Carriers Fabricated by the FDM Printing Process

Nada Ben Hariz[1,2](✉), Atef Boulila[2], and Mahfoudh Ayadi[3]

[1] Production and Energy Laboratory (LMPE), National Engineering School of Tunis, University of Tunis El Manar, BP 37, Belvédère, 1200 Tunis, Tunisia
nada.benhariz@enit.utm.tn

[2] National Institute of Applied Sciences and Technology, University of Carthage, BP676, 1080 Tunis Cedex, Tunisia
atef.boulila@insat.ucar.tn

[3] National Engineering School of Bizerte, University of Carthage, Menzel Abderrahmen, BP667035 Bizerte, Tunisia

Abstract. The Fused Deposition Modeling (FDM) process is one of the most widespread additive manufacturing technologies today. However, it still suffers from a great problem of anticipation of the quality of the manufactured product. Indeed, the quality of the output product is a key parameter in the acceptability of the end user. Many fields such as aeronautics, biomedical, food, pharmaceutical require many strict conditions for the acceptability of the product. In biological fields, and especially in the pharmaceutical field, the 3D printed product must meet the standards and requirements of the field to be accepted. In this context, our paper aims at analyzing the influence of some input parameters on the geometrical and topographical quality of samples dedicated to a biological application called bio-carrier. These bio-carriers are small structures used to capture phages when they are present at low titers. The 3D printed bio-carrier has also been used to enrich in vitro preparations for priming phage cultures to rapidly achieve high titers suitable for phage therapy applications. Phage therapy is used as a potential substitute for chemical antibacterial therapy.

A Design Of Experience was adopted, in which the size, thickness and filling rate of the bio-carrier were chosen as input parameters. The analysis of this design showed that the sample size is the key parameter that can improve or reduce the geometric effect. Increasing the sample size can reduce the geometric defect. The sample topography is influenced by the fill rate which is the most important factor.

Keywords: Additive manufacturing · Fused Deposition Modeling (FDM) · Polylactic Acid (PLA) · Geometric Defect · Topography Defect

1 Introduction

The Additive manufacturing (AM) is a process that creates products with complex geometry using a variety of advanced materials and techniques.

T. Mabrouki et al. (Eds.): AIAM 2023, LNME, pp. 44–51, 2024.
https://doi.org/10.1007/978-3-031-47784-3_6

Fused deposition modeling (FDM) is an additive manufacturing technique that produces three-dimensional objects by building successive layers of raw material (Rahim, Abdullah, and Md Akil 2019). Samples are produced from a computer-aided design (CAD) drawing, allowing the manufacturer to easily make changes or adapt the desired product.

Many printing parameters can be changed during the preparation of the print pattern, such as print speed, nozzle temperature, wall thickness, fill rate, fill strategy, nozzle size, build plate temperature, etc. (Fernandes et al., n.d.). Poorly defined parameters can lead to several defects in the printed parts, namely poor surface quality, poor dimensional accuracy (Ben Hariz, Boulila, and Ayadi, n.d.), reduced mechanical strength, the presence of many empty print areas in the parts that hinder or even prevent their use (Ben Hariz et al. 2022).

The FDM process has become a leader in AM techniques, thanks to its various fields of application (Ben Hariz et al. 2022) (Negi, Dhiman, and Sharma 2013). One of its new applications is biological and more particularly in the field of phage therapy (Li et al. 2018).

Today, due to the extensive use of antibiotics, bacteria have developed resistance to them, decreasing their effectiveness (Housby and Mann 2009). Since their discovery in the early 20th century, bacteriophages, or natural predators of bacteria, have been used effectively in many scientific studies to treat infections (Lariyah et al. 2016). For researchers and bioengineers, the laborious separation of phages from their natural environment and the enrichment of their cultures to achieve the highest titers for effective antibacterial potential still pose challenges (Barwal and Chaudhary 2014).

For this reason, a new structure called a bio-carrier has been fabricated (Sfetsas, Patsatzis, and Chioti 2021). Nowadays, 3D printing with a very complicated shape, many dimensions and low cost produces this bio-carrier (Zhu et al. 2021).

In the same context, the idea to study the ability of a 3D printed bio-carrier structure using polylactic acid (PLA) to trap target phages (Bouchart et al. 2020). Similarly, our study analyzes the behavior of these printed parameters by evaluating geometric and topography defects (Reddy et al., n.d.).

In this paper, a series of biological carrier structures of different dimensions and parameters were studied to examine geometric and topographical defects, which will then be compared to biological results obtained for the same specimens.

2 Materials and Methods

2.1 Experimental Methods and Materials

In this paper, a hexagonal sample was chosen, as showed the digital camera photo in the Fig. 1, to be used in phage capture for phage therapy. According to the biological instructions, the bio-carrier should have maximum external surface area and small dimensions to fit in the beaker. In addition, the outer surface of the bio-carrier must have a minimum of empty area to ensure that the biofilm covers all surfaces (Barwal and Chaudhary 2014). Therefore, it was decided in this work to test two bi-cavity sizes, two thicknesses and two filling rates.

Fig. 1. Hexagonal form of bio-carrier

All samples are designed using SolidWorks software and printed using Creality CR5-Pro printer using FDM process and white PLA filament. In order to study the influence of the same parameters on geometric and topographical defects, three factors were chosen: dimension, thickness, and fill rate. The other parameters were kept at the values recommended in previous work. Table 1 summarizes the input parameters used to perform the bio-carriers.

Table 1. Printing Process Parameters

Parameters	Unit	Value
Wall thickness	mm	1.32
Top/bottom thickness	mm	0.8
Infill pattern	–	linear
Raster angle	°	45°
Build plate temperature	°C	60

2.2 Design Of Experience (DOE).

In this work, the Taguchi method was used. The Taguchi use the Design Of Experiment (DoE) to reduce geometric defect and topography of the bio-carrier structure (Wicaksono and Felix Krisna Aji Nugraha 2022). In our DoE, we have three factors and two levels as presented in the Table 2.

The Table 3 presents the complete DoE on the Taguchi L8 Orthogonal Array (OA).

The Fig. 2 showed all 3D printed specimen's photos of the DoE taken with a digital camera.

In this work, geometric defects will be evaluated based on measuring the 120° angle for each hexagonal shape of the samples and calculating the deviation between the programmed angle and the actual printed angle.

Table 2. Factors and Levels of DoE

Levels	Dimension (mm)	Thickness (mm)	Infill density (%)
1	30	5	20
2	15	3	80

Table 3. DoE on the Taguchi L8 orthogonal array

Levels	Dimension (mm)	Thickness (mm)	Infill density (%)
1	30	5	80
2	30	5	20
3	30	3	80
4	30	3	20
5	15	5	80
6	15	5	20
7	15	3	80
8	15	3	20

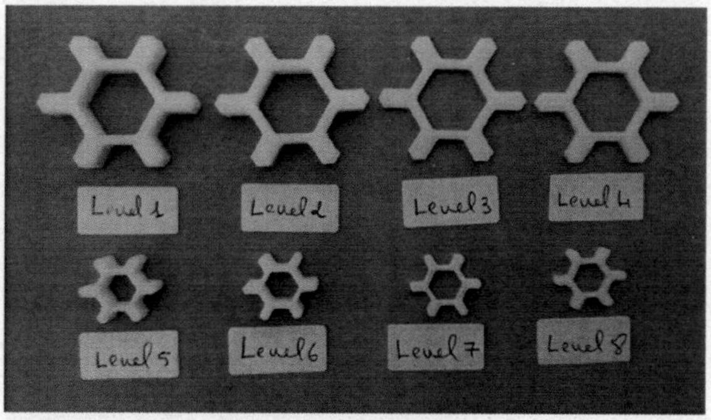

Fig. 2. Printed Specimens

All angles are measured with a handheld protractor to an accuracy of 1/10 mm. The topography of our bio-carrier structure is evaluated by microscopic observations using an optical microscope with a ZuZi 5Mp digital camera to visualize the microstructure of the samples. The visualization is done at 4 different points of each 3D printed bio-carrier structure as presented in Fig. 3.

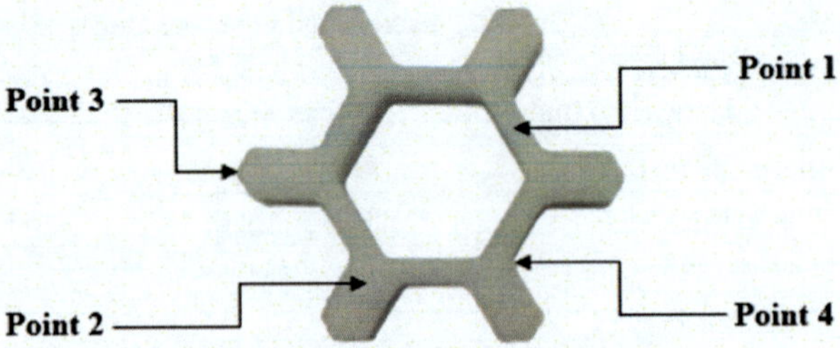

Fig. 3. Visualization point

3 Results and Discussion

All experimental results will be discussed for the biological problem. In future work, we plan to compare these results with the biological results obtained with the same specimens. The ability to increase the number of phages is the most important criteria.

3.1 Geometric Defect

Table 4 shows the 120° difference between the programmed angle and the printed angle for each specimen.

Table 4. Angles Gap

Levels	Dimension (mm)	Thickness (mm)	Infill density (%)	Angle Gap (°)
1	30	5	80	0.89
2	30	5	20	1.69
3	30	3	80	1.32
4	30	3	20	0.92
5	15	5	80	3.45
6	15	5	20	4.03
7	15	3	80	2.71
8	15	3	20	3.21

After analyzing the results and interpreting them using the MINITAB software, as shown in Fig. 4, we find that the highest slope of the average effect chart is in the column of the circumscribed circle dimension. Therefore, the most important factor that can influence the geometric defect is the dimension of our product. ANOVA results for our DoE ensures our approach that dimension impacts more on geometry defect (87.37%) next thickness (4.28%) and later infill density (2.60%).

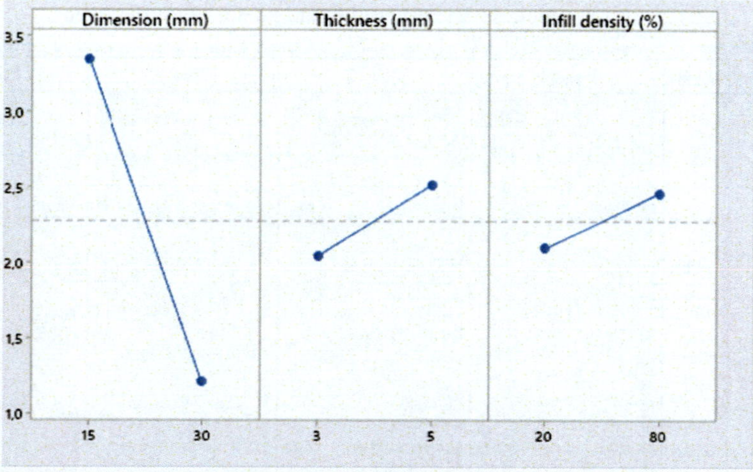

Fig. 4. Average effect chart

3.2 Topography Defect

Table 5 shows the optical microscope with a ZuZi 5Mp digital camera visualization of each specimen at 4 different points at 100x magnification (all specimens are visualized under the same conditions and position).

The distance between two layers and any void areas are evaluated using ImageJ software.

After analyzing all the microscopic visualizations, we find that the fill density is the main factor influencing the topography of the sample. The worst sample is located at level 8 with 15 mm in dimension, 3 mm in thickness, and a fill density of 20%. This sample has many defects and void areas larger than 0.05 mm, which can cause many problems for phage recovery. The best samples with respect to topography defects are the samples in levels 1, 2, 3, and 4. The void area in this sample is half as small as in the other samples.

A high fill density can improve topography, but it also increases product weight. In the case of phage therapy, the weight reduces the stirring behavior of our bio-carrier. For specimens with small dimension the infill density doesn't have any influence in the topography because the dimension between 2 layers is smaller than the layer length.

To optimize the topography defect, we can use the maximum size and maximum filling density, so it is recommended to choose the largest size that can be put in the beaker and can completely blend into the biofilm.

Table 5. Microscope Visualization for each specimen

Levels	Point 1	Point 2	Point 3	Point 4
1			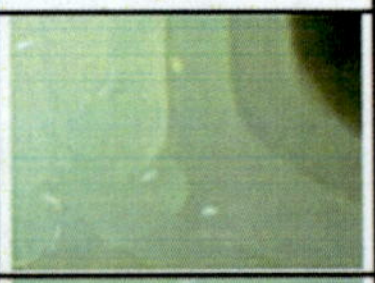	
2				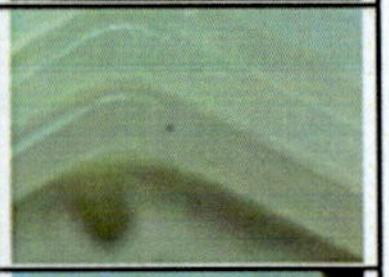
3	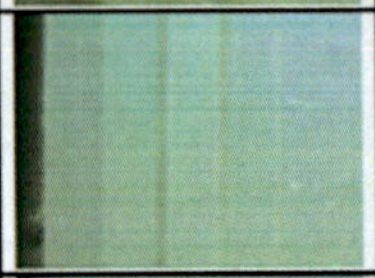			
4			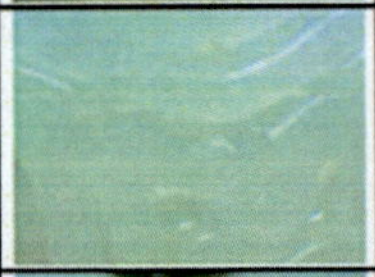	
5				
6		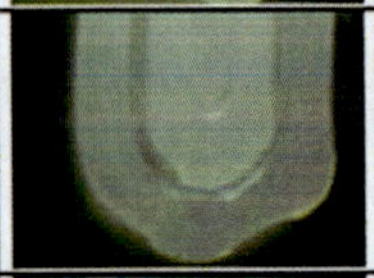	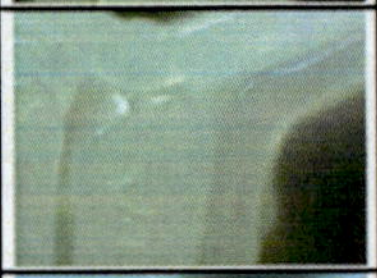	
7				
8		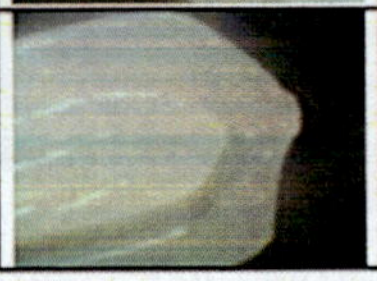	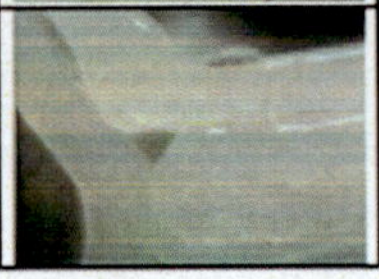	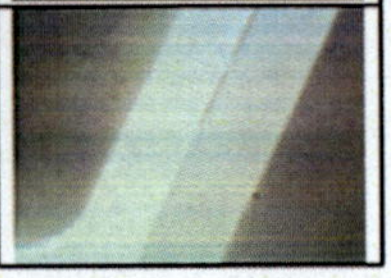

4 Conclusion and Perspectives

Current research presents an overview of the use of additive manufacturing (FDM) in biological fields to improve and facilitate phage therapy.

Bio-carrier size is a key parameter in the geometry and topography defect. When the product size is reduced, the print quality decreases significantly.

These results will be compared to a biological study to determine the most appropriate level for phage extraction.

References

Barwal, A., Chaudhary, R.: To study the performance of biocarriers in moving bed biofilm reactor (MBBR) technology and kinetics of biofilm for retrofitting the existing aerobic treatment systems: a review. Rev. Environ. Sci. Bio/Technol. **13**(3), 285–299 (2014). https://doi.org/10.1007/s11157-014-9333-7

Ben Hariz, N., Boulila, A., Ayadi, M., Fattouch, S.: Evaluation of a complete characterization of polylactic acid used in additive manufacturing: biomedical application. Series Biomech. **36**(3) (2022). https://doi.org/10.7546/SB.02.03.2022

Ben Hariz, N., Boulila, A., Ayadi, M.: The Influence of 3D Printing Process Parameters in the Dimension Accuracy, Roughness, and Weight

Ben Hariz, N., Boulila, A., Ayadi, M., Fattouch, S.: Exploitation of the Additive Manufacturing Process to Produce Replacement Knee Exo-Prostheses, no. 2 (September): 14 (2022)

Bouchart, F., et al.: 3D Printed bioceramic for phage therapy against bone nosocomial infections. Mater. Sci. Eng., C **111**, 110840 (2020). https://doi.org/10.1016/j.msec.2020.110840

Fernandes, J., Deus, A.M., Reis, L., Vaz, M.F., Marco, A., Leite. N.D. :'Study of the Influence of 3D Printing Parameters on the Mechanical Properties of Pla', 7

Housby, J.N., Mann, N.H.: Phage therapy. Drug Discovery Today **14**(11–12), 536–540 (2009). https://doi.org/10.1016/j.drudis.2009.03.006

Lariyah, M.S., et al.: Application of Moving Bed Biofilm Reactor (MBBR) and Integrated Fixed Activated Sludge (IFAS) for biological river water purification system: a short review. IOP Conf. Series: Earth Environ. Sci. **32**, 012005 (2016). https://doi.org/10.1088/1755-1315/32/1/012005

Li, Z., Rathore, A.S., Song, C., Wei, S., Wang, Y., Xu, W.: 'PrinTracker: fingerprinting 3d printers using commodity scanners. In: Proceedings of the 2018 ACM SIGSAC Conference on Computer and Communications Security, Toronto Canada, pp. 1306–23. ACM (2018). https://doi.org/10.1145/3243734.3243735

Negi, S., Dhiman, S., Sharma, R.K.: Basics, applications and future of additive manufacturing technologies: a review. J. Manuf. Technol. Res. **5**(1/2), 75 (2013)

Rahim, T.N., Tuan, A., Abdullah, A.M., Akil, H.M.: Recent developments in fused deposition modeling-based 3D printing of polymers and their composites. Polym. Rev. **59**(4), 589–624 (2019). https://doi.org/10.1080/15583724.2019.1597883

Reddy, V.V., Flys, O., Krishna, A.V., Rosen, B.-G.: n.d. 'Topography Characterization of Fused Deposition Modelling Surfaces'

Sfetsas, T., Patsatzis, S., Chioti, A.: A review of 3D printing techniques for bio-carrier fabrication. J. Clean. Prod. **318**, 128469 (2021). https://doi.org/10.1016/j.jclepro.2021.128469

Wicaksono, M.B., Nugraha, F.K.A.: Optimization of 3D printing parameters using the taguchi method to improve dimensional precision. Jurnal Teknologi, pp. 13–18 (2022). https://doi.org/10.35134/jitekin.v12i2.72

Zhu, Y., et al.: 3D Printing biomimetic materials and structures for biomedical applications. Bio-Design Manuf. **4**(2), 405–428 (2021). https://doi.org/10.1007/s42242-020-00117-0

Determination of Mechanical Parameters of 3D-Printed Parts: Experimental and Numerical Analyses

Sami Chatti[1](✉) and Amina Remadi[2]

[1] Higher Institute of Applied Science and Technology of Sousse, Tahar Ben Achour, 4003 Sousse, Tunisia
s.chatti.issat@gmail.com

[2] National Engineering School of Sousse ENISo, Mechanical Laboratory of Sousse LMS, University of Sousse, Sousse, Tunisia

Abstract. Additive manufacturing process allows obtaining structured light parts widely desired by designers. The mechanical behavior of such materials is anisotropic and their mechanical parameters have to be precisely determined in order to obtain reliable numerical simulations. In this paper, the mechanical parameters of a grid pattern with an infill density of 60% are experimentally determined. The elasticity matrix is determined by considering a cubic anisotropy. In addition, the Hill48 criterion is adopted and rewritten in the case of the cubic anisotropy. For this end, the tensile and the shear tests are used in order to obtain the mechanical parameters in both elastic and plastic domains. For the sake of validation of the obtained parameters, flat-ring parts are additively manufactured by considering two raster orientations: 0°/90° and −45°/45°. These parts are subjected to the compression test and the applied load vs. the punch stroke is experimentally obtained showing the influence of the printing parameters on the mechanical responses. The effect of the raster orientation is then clearly observed showing that the raster orientation of −45°/45° leads to more resistant parts. Finally, the obtained numerical results are compared with experiment. It is found that the numerical simulations fit very well the experimental results.

Keywords: Additive manufacturing · tensile test · orthotropic parameters · Hill criterion

1 Introduction

The adoption of the additive manufacturing (AM) has been continuously growing [1] undoubtedly due to the accessibility of inexpensive 3D-printing devices becoming affordable even for home use. Lightweight design by the AM is a branch of mechanical engineering that is increasingly used in many industries, as they provide reduced mass and more efficient parts. In addition, AM allows manufacturing complex shapes which cannot be obtained by conventional manufacturing techniques. The determination of the mechanical properties is of crucial importance in order to obtain valuable numerical

T. Mabrouki et al. (Eds.): AIAM 2023, LNME, pp. 52–59, 2024.
https://doi.org/10.1007/978-3-031-47784-3_7

simulations. However, such materials reveal complex microstructure which is difficult to be modeled. [2, 3, 4] have experimentally investigated on the effect of different process parameters as infill pattern, infill density, layer thickness and printing orientation, on mechanical strength of Poly-lactic Acid (PLA) material. It has been established that the mechanical behavior of structured specimens is anisotropic justified by their behavior under tensile tests when considering different raster orientations [5, 6]. [3] assumed a transversely isotropic material for the elasticity model and Tsai-Hill strength criterion under plane stress state.

Numerical simulations of parts obtained by the AM are very expensive in time consuming. Consequently, replacing the discontinuous geometry (lattices) by its equivalent continuous ones in terms of mechanical behavior leads to reasonable numerical simulations duration. Thus, adopting a suitable behavior law and using reliable mechanical parameters is of crucial importance to success numerical simulations. It is important to highlight that there are few studies on the establishment of the anisotropic parameters specially when considering low infill densities. The scope of this paper is to experimentally determine the anisotropic parameters for the grid patterns with a relatively low density of 60% by considering the tensile and shear tests. A flat-ring parts, obtained by AM and subjected to compression test, are investigated. Experimental and numerical simulations are conducted showing the relevance of the adopted mechanical and numerical modeling.

2 Experimental Results

The used filament is a biodegradable polymer (PLA) printable on Raise 3D machine, with a filament diameter of 1.75 mm. The used tensile machine is an MTS insight at ISSAT Sousse. Figure 1 gives the tensile test of the PLA filament (solid line) and two specimens made of grid infill pattern with density of 100% and two raster orientations (dashed line for 0°/90° and −45°/45°). Figure 2.a) displays the tensile ASTM D638 specimen as reported by Giulia et al. [7]. It is clear that the infill orientation has not a significant effect on the mechanical response except for the ultimate strain at fracture. Also, AM with the density of 100% leads to a slight softening when compared with the PLA filament stiffness. Table 1 gives the obtained mechanical properties of the filament and the specimens made of AM. Notice that 3 samples were used for every test.

Figures 2b) and 2b) show the fractured specimens for raster orientations of 0°/90° and −45°/45° respectively. These figures show that the direction of fractures is intimately dependent on the raster orientation. Figure 2d) shows the same fracture orientation as observed in Fig. 2c) for the infill density of 60% and raster orientations of −45°/45°. These observations were recently obtained by numerical simulations in [8]. Next, only the tensile specimens made of the grid infill pattern and the infill density of 60% are considered in order to investigate its effect on the mechanical parameters when considering two infill orientations. The displacement is measured by an extensometer.

Figure 3 gives the stress-strain response of the specimens made of the grid infill pattern with density of 60%. It is clear that the 3D-print orientations of 0°/90° is slightly more resistant than that of −45°/45° orientations. However, this raster orientation leads to higher ultimate strain at fracture.

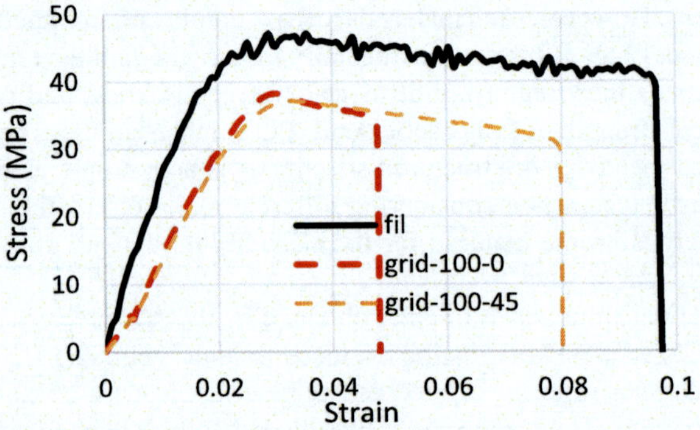

Fig. 1. Tensile test on the PLA filament and the grid infill pattern

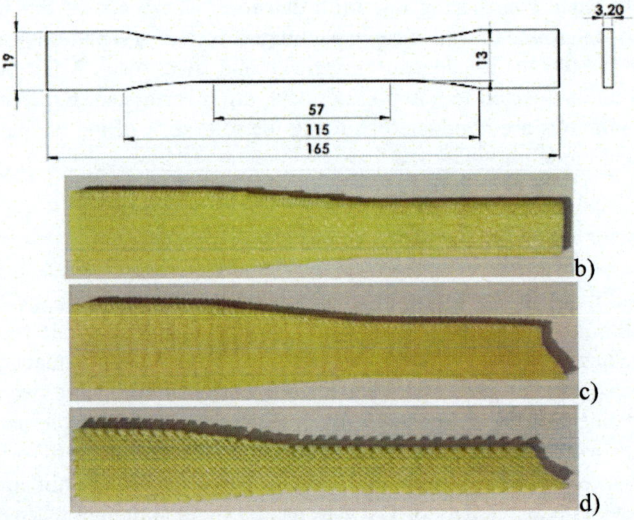

Fig. 2. Tensile specimen; a) Dimensions of the tensile ASTM D638 specimen b) fractured specimen density of 100% and orientation 0°/90°; c) fractured specimen density of 100% and orientations −45°/45°; d) fractured specimen density of 60% and orientations −45°/45°

Table 1. Mechanical properties

	Elastic modulus (MPa)	Yield stress (MPa)	Ultimate strain
Filament °	2405	41	9.6%
Grid (100%)	1420	36	4.5% / 8%

A shear device was designed and adapted on the tensile machine. The specimens, shown in Fig. 4, are of dimensions of 30 × 10 × 5 mm^3. In Fig. 5 is plotted the shear

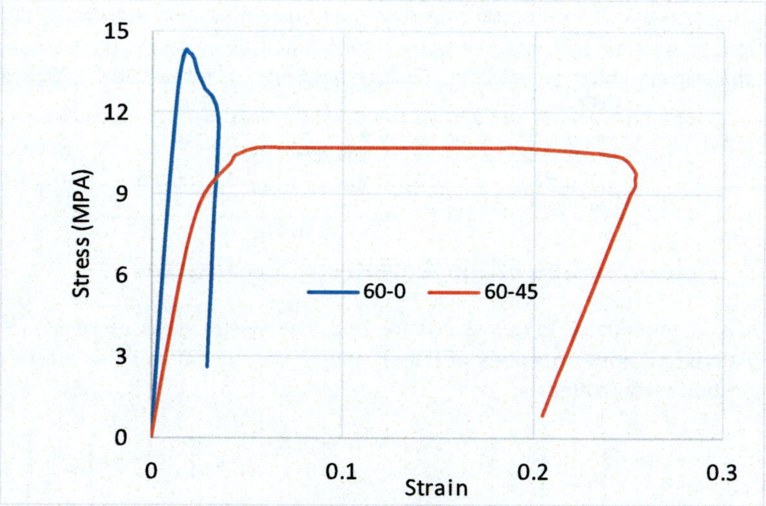

Fig. 3. Tensile test

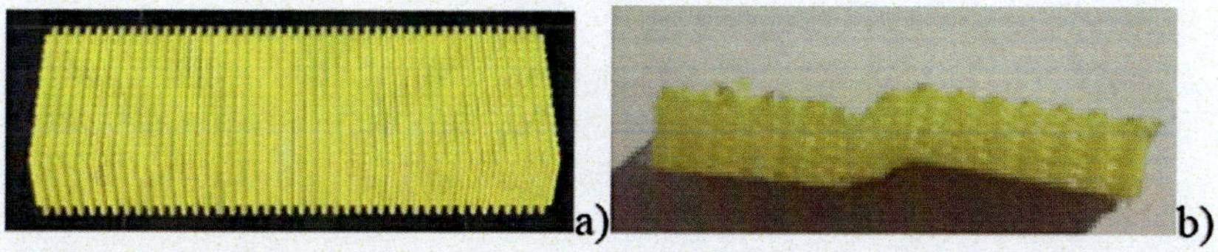

Fig. 4. Shear test; a) Specimen with the grid infill pattern, density of 60% and orientations of 0°/90°; b) Deformed specimen

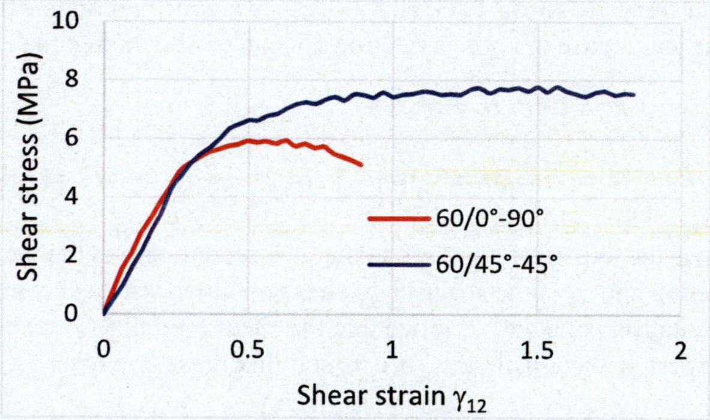

Fig. 5. Shear test

curves. This figure shows that the effect of the raster orientation is not very important on the values of the mechanical parameters as displayed in Table 2.

Table 2. Mechanical properties of the grid infill pattern with density of 60%

Orientation	Elastic modulus (MPa)	Yield stress (MPa)	Shear modulus (MPa)	Yield shear (MPa)
0°/90°	774 ± 44	14.3 ± 0.4	16.7 ± 0.5	5.7 ± 0.4
45°/-45°	403 ± 88	10.6 ± 0.3	15.3 ± 0.6	6.2 ± 0.5

3 Determination of the Anisotropy Coefficients

It is of importance to notice that the behavior of structures made with the grid infill pattern and print directions of 0°/90° can be considered as cubic anisotropy with the elasticity law written as:

$$\begin{Bmatrix} \varepsilon_{11} \\ \varepsilon_{22} \\ \varepsilon_{33} \\ \gamma_{12} \\ \gamma_{13} \\ \gamma_{23} \end{Bmatrix} = \begin{bmatrix} 1/E & -\nu/E & -\nu/E & 0 & 0 & 0 \\ -\nu/E & 1/E & -\nu/E & 0 & 0 & 0 \\ -\nu/E & -\nu/E & 1/E & 0 & 0 & 0 \\ 0 & 0 & 0 & 1/G & 0 & 0 \\ 0 & 0 & 0 & 0 & 1/G & 0 \\ 0 & 0 & 0 & 0 & 0 & 1/G \end{bmatrix} \begin{Bmatrix} \sigma_{11} \\ \sigma_{22} \\ \sigma_{33} \\ \sigma_{12} \\ \sigma_{13} \\ \sigma_{23} \end{Bmatrix} \quad (1)$$

where E stands for the Young's modulus, ν is the Poisson's ratio ($\nu = 0.3$), G is the shear modulus, ε_{ii} are the normal strains, γ_{ij} are the shear strain and σ_{ij} are the stress tensor components. With respect to the orthotropic axes, the following Hill48 criterion is adopted:

$$\begin{aligned} \bar{\sigma}^2 &= F(\sigma_{22} - \sigma_{33})^2 + G(\sigma_{33} - \sigma_{11})^2 \\ &+ H(\sigma_{11} - \sigma_{22})^2 + 2L\sigma_{23}^2 + 2M\sigma_{31}^2 + 2N\sigma_{12}^2 \end{aligned} \quad (2)$$

where $\bar{\sigma}$ is the yield stress, F, G, H, L, M and N are the parameters of the Hill48 criterion. In the particular case of cubic anisotropy, it can be shown that the Hill48 parameters are:

$$\begin{aligned} &F = G = H = 0.5 \\ &N = L = M = \frac{1}{2}\left(\frac{\sigma_0}{\sigma_c}\right)^2 = 2\left(\frac{\sigma_0}{\sigma_{45}}\right)^2 - 1/2 = 3.1 \pm 0.3 \end{aligned} \quad (3)$$

where σ_0 is the tensile yield stress in the 0° direction, σ_{45} is the tensile yield stress in the 45° direction and σ_c is the shear yield stress obtained with a specimen made of AM with print orientations of 0°/90°. Notice that the shear modulus G can be determined through the shear test or the tensile test. It is found that these two tests give close results.

4 Simulation of a Flat-Ring Part Under Compression

A flat ring part, with dimensions given in Fig. 6a), is experimentally and numerically investigated. The grid infill pattern with density of 60% and the two infill orientations of 0°/90° and −45°/45° are considered. Figure 6b) gives the set-up of the compression test. The part is positioned between a support and a punch. A friction coefficient of 0.5, between the part and the tools, is experimentally determined with an inclined plane method. The mechanical properties displayed in Table 2 are used for the subsequent simulations. Notice that the plastic behavior is assumed to be perfect (without hardening and damage). After some meshing optimizations, the part is meshed with 2802 of plane stress elements as displayed in Fig. 7. Figure 8 shows the experimental and the numerical deformed flat-rings made by the AM for a punch stroke of 15 mm. This figure clearly shows that the predicted shapes by numerical simulations are qualitatively close to the experimental forms. It seems that the deformation depends on the alignment of the loading with the raster orientation. Once again, it is observed that the print directions of −45°/45° provides a resistant part without damage when compared with the part printed through the directions of 0°/90°.

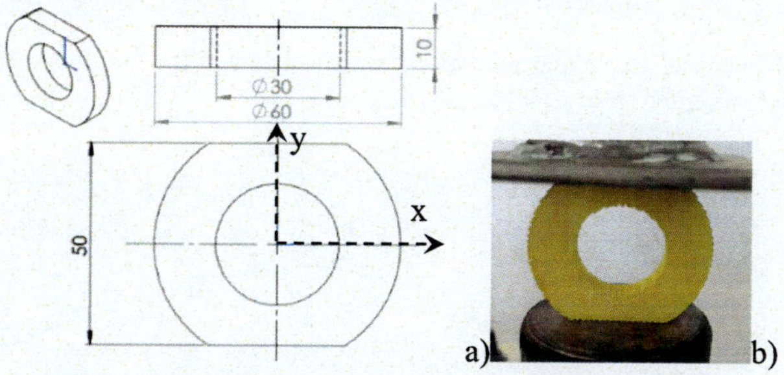

Fig. 6. Flat-ring part a) dimensions b) compression test

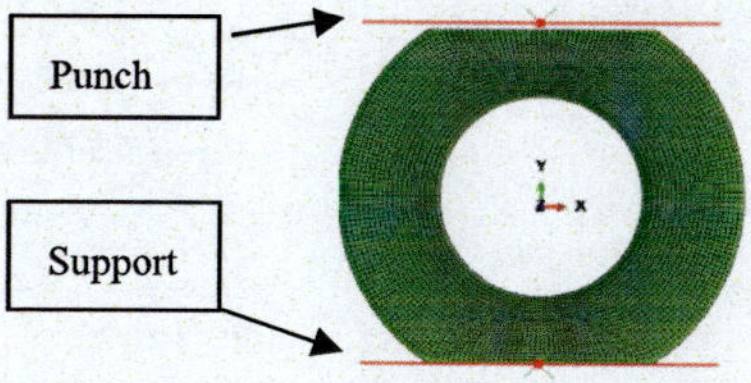

Fig. 7. Boundaries conditions and meshed part

Figure 9 displays the punch load vs. the punch displacement for the two considered print-directions. It can be seen that the curve with the infill orientations of −45°/45° contains a plateau, which makes the part more resistant than the flat-ring made with the raster orientations of 0°/90°. In addition, the comparison with the experimental data shows a good agreement. It can be concluded that:

- the adopted cubic anisotropy behavior is justified;
- the mechanical parameters, experimentally obtained, are reliable;
- the damaged regions in the deformed flat-ring with print orientations of 0/90° are located in relatively small areas. Consequently, its effect on the loading curve is not significant, as it was observed in Fig. 9.

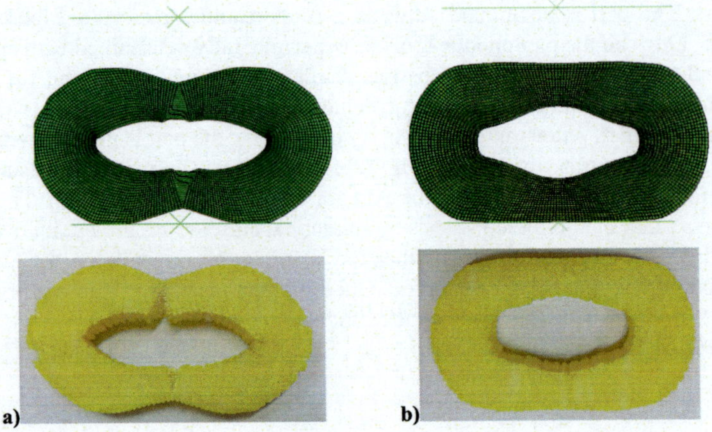

Fig. 8. Numerical and experimental deformed flat-rings for the punch displacement of 15 mm and pattern orientation of a) 0°/90°; b) −45°/45°

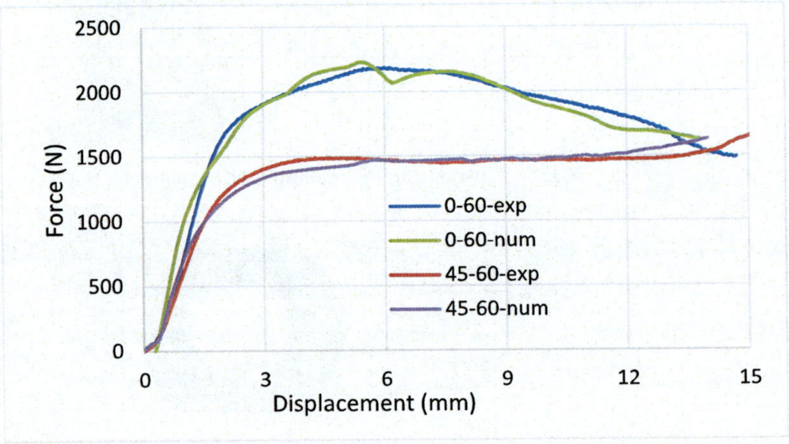

Fig. 9. Punch force vs. punch stroke

5 Conclusion

In this study, the grid infill pattern with density of 60% and two printing orientations of 0°/90° and −45°/45 were used to manufacture specimens in order to determine the mechanical parameters by using tensile and shear tests. A cubic anisotropy behavior was considered for both the elastic matrix and the Hill48 criteria. It was established that the tensile and the shear tests applied for 3D-printed specimens with 0°/90° and −45°/45 orientations were sufficient to acquire elastic and plastic mechanical parameters of such material. Experimental tests and numerical simulations were investigated on a flat-ring subjected to compression test. The comparison between the numerical and the experimental data showed a good agreement which validate the assumed anisotropic behavior law. Furthermore, it was found out that the part with infill orientation of –45°/45° allows large deformation without damage as it was found in the tensile test. As a conclusion, in order to obtain more resistant parts, it is recommended that the applied loads had not to be aligned to the raster orientation. This study can be extended for other infill patterns with various infill densities.

References

Joshi, S.C., Sheikh, A.A.: 3D printing in aerospace and its long-term sustainability. Virtual Phys. Protot. **10**, 175–185 (2015)

Ei, C., et al.: Investigation on influence of infill pattern and layer thickness on mechanical strength of PLA material in 3D printing technology. J. Eng. Sci. Res. **3**(2), 27–37 (2019)

Zhao, Y., Chen, Y., Zhou, Y.: Novel mechanical models of tensile strength and elastic property of FDM AM PLA materials: experimental and theoretical analyses. Mater. Des. **181**, 108089 (2019)

Camposeco-Negrete, C., Varela-Soriano, J., Rojas-Carreón, J.J.: The effects of printing parameters on quality, strength, mass, and processing time of polylactic acid specimens produced by additive manufacturing. Progress Add. Manuf. **6**, 821–840 (2021)

Letcher, T., Waytashek, M.: Material property testing of 3D-printed specimen in pla on an entry-level 3D printer. 2014; A:ASME (2014)

Song, Y., Li, Y., Song, W., Yee, K., Lee, K.-Y., Tagarielli, V.L.: Measurements of the mechanical response of unidirectional 3D printed PLA. Mater. Des. **123**(5), 154–164 (2017)

Giulia, et al.: Comprehensive characterization of mechanical and physical properties of PLA structures printed by FFF-3D-printing process in different directions. Progress Add. Manuf. **7**, 1111–1122 (2022). https://doi.org/10.1007/s40964-022-00285-8

Chatti, S., et al.: Experimental investigations and Numerical simulations of lattice structures in additive manufacturing. In: JSTMM 2022, 20–21 December, Monastir– Tunisia (2022)

3D Printed Educational Robotic Arm

Mohamed Slim[1(✉)] and Mohamed Ali Terres[1,2]

[1] Laboratory of Mechanics, Materials and Processes (LMMP), University of Tunis, National High School of Engineering of Tunis (ENSIT), 5, Avenue Taha Hussein, Montfleury, 1008, Tunis, Tunisia
mohamed.slim@ensit.u-tunis.tn

[2] University of Sousse, High Institute of Sciences and Applied Technology of Sousse (ISSATSo), Cité Taffala, Ibn Khaldoun, 4003 Sousse, Tunisia

Abstract. This paper presents an exploration concerning the employment of the additive manufacturing process in the fabrication of educational robot prototypes that mimic the technical qualities and operating characteristics of industrial robotic arms. Specifically, we present the 3D printing process of an open-source model called the EEZYbotARM MK2, using Fused Deposition Modeling (FDM). This model shares the kinematic structure of the industrial ABB IRB 460 robot. The 3D printed is proposed as an educational robot for the Higher Institute of Applied Sciences and Technology of Sousse (ISSAT-So). The 3D printing process involves the use of PLA filament and the RAISE3D N2 printer. We present the proposal printing parameters applied to achieve an efficient quality of the printed parts, within a minimum amount of printing time. Finally, we present the assembled EEZYbotARM MK2 and its potential applications for educational purposes in the field of robotics. The results demonstrate the potential of 3D printing in the development of educational robot prototypes, and the importance of open-source models for the circularity and the availability of robotics education.

Keywords: Additive manufacturing · 3D printer · fused deposition modeling · robotics · educational robot manipulators

1 Introduction

Robotics is an interdisciplinary field integrating mechanics, electronics, and computer science (Thrun et al., 2005). Robots are typically composed of a mechanical structure, electronic control unit, and communication interface that enable interaction with users and ensure effective operation in their environment.

Even if the main constitution of robots remains the same for different robots. They have never ceased to develop, as this field is always considered a subject of innovation and development (Rosheim, 1994). This makes teaching robotics in universities of applied sciences and technologies a key issue.

Teaching robotics is a pedagogical challenge in many universities (Alimisis, 2013; Altin & Pedaste, 2013). The theoretical aspect of robotics is usually mathematically complex, since it may seem vague and very abstract to most students, a real robotic

T. Mabrouki et al. (Eds.): AIAM 2023, LNME, pp. 60–69, 2024.
https://doi.org/10.1007/978-3-031-47784-3_8

model is essential to ensure a good teaching in this field. On the other hand, existing models of robots are usually very expensive, mainly robotic arms. 3D printing prototypes are a potential alternative to real models in robotics education. In fact, a 3D printed model has all the technical qualities and operating characteristics of the original model.

Several researchers have adopted the implementation of the 3D printing process, exposing the potentials of such technique in the development of low cost and yet sufficient educational robots. For instance, Gonzalez-Gomez et al. (2012), proposed a low cost 3D printable mobile robot for educational purposes. It allows students to learn programming as well as to modify the robot's chassis and create new custom parts. Chavdarov et al. (2019) developed an educational redundant 3D printed robot. Telegenov et al. (2015) proposed an open-source low-cost three-finger robotic gripper platform for research and educational purposes.

This paper focuses on the 3D printing process of an open-source robot known as EEZYbotARM MK2 (Franciscone, 2016).

It is a prototype of the industrial ABB IRB-460 robot. The fabricated model is proposed as an educational robot material for the Higher Institute of Applied Sciences and Technology of Sousse (ISSAT-So).

The research aspect of this paper lies in demonstrating the feasibility of using this technology for educational purposes in the field of robotics. 3D printed robot prototypes can enhance students' understanding of theoretical concepts, through a practical approach.

The paper is organized as follows: In Sect. 2, we briefly introduce the real industrial robot ABB IRB 460 followed by a detailed presentation of the prototype design, EEZYbotARM MK2. Section 3, exposes the specifications of printing the robotic arm using the 3D printer RAISE3D-N2 available at ISSAT-So. While, exhibiting the slicing of each printable part of the robot as well as the final model of the robot after assembly. The final section is dedicated to conclusion.

2 Prototype's Design

The aimed robot's prototype emulates the kinematic structure of the industrial ABB IRB460 robot manipulator. The following subsections will provide; 1. A brief description of the industrial ABB IRB460, 2. Parts design of the ABB IRB460' prototype, 3. The 3D printing process of the prototype's printable components and finally the modeling and control of the developed prototype.

The ABB IRB460 shown in Fig. 1, is a four-axis industrial robot, designed and manufactured by ABB. It is a versatile and high-performance robot that is used mainly in industrial applications, such as palletizing, de-palletizing, and material handling. Characterized by a payload capacity of up to 110 kg and a reach of up to 2.4 m, the IRB460 is capable of handling heavy and bulky objects with great precision and speed (*IRB 460 | ABB Robotics*, n.d.).

The kinematic linkage of the developed prototype could be decomposed into three main parts; the base, the parallelogram linkage mechanism and the gripper as presented in Fig. 2. The base is bound to a specific platform, the parallelogram linkage mechanism is adopted to remotely control the movement of the gripper according to the two

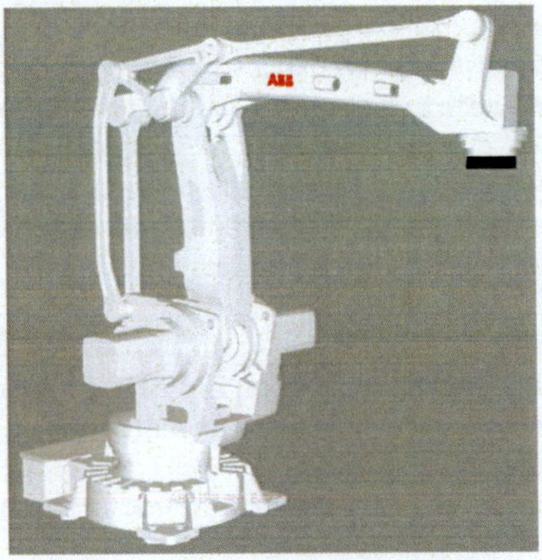

Fig. 1. ABB IRB 460 industrial robot

servomotors fixed to the base, while maintaining a constant orientation of the gripper (parallel to the floor).

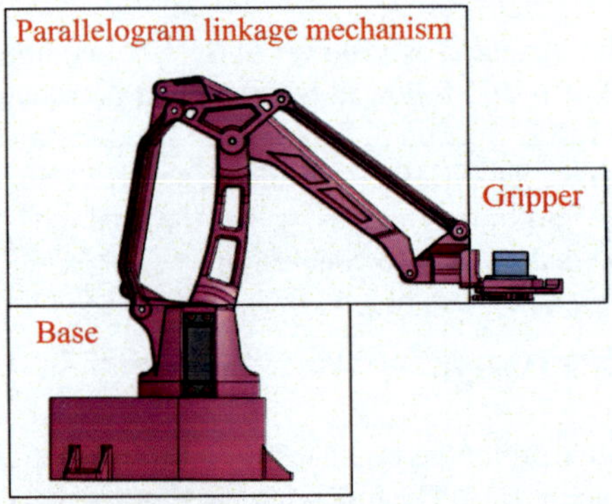

Fig. 2. Robot's main parts

2.1 Base Parts

The main base of the robot is composed of 5 printable parts as presented by exploded view in Fig. 3.

2.2 Parallelogram Linkage Parts

The parallelogram mechanism adopted in this prototype is constructed of 8 components as illustrated in Fig. 4.

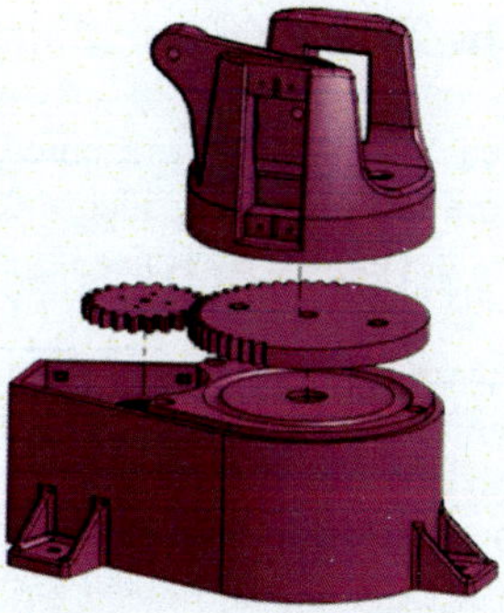

Fig. 3. Exploded view of the robot's base

2.3 The Gripper Parts

The gripper is the end-effector tool of the robot attached to its end, which allows the robot to grasp and manipulate objects. The gripper's role is to securely grasp and release objects, allowing the robot to move and manipulate them in various ways. In our case, the gripper consists of 5 printable parts illustrated in purple in Fig. 5.

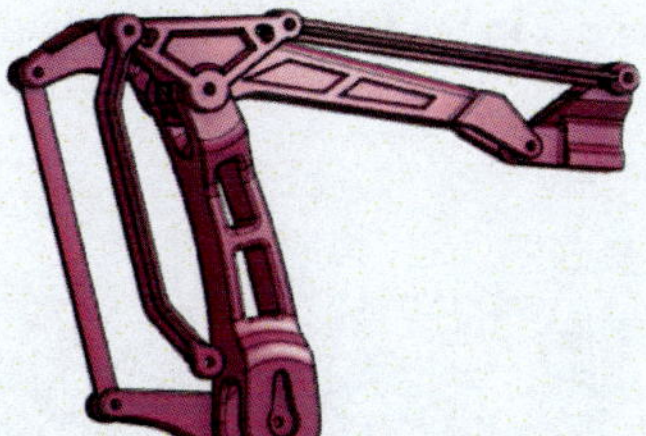

Fig. 4. Parallelogram linkage of the robot

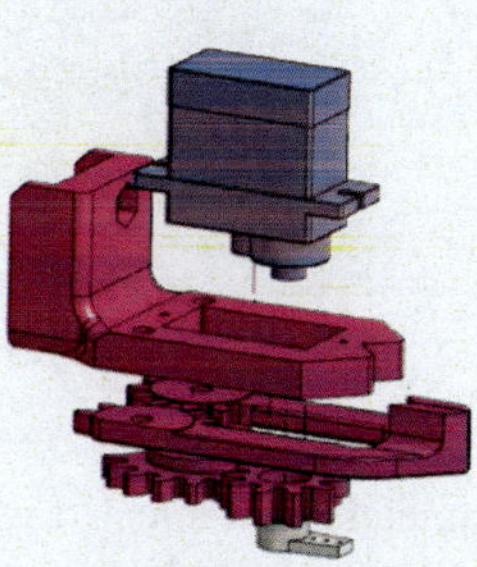

Fig. 5. Exploded view of the robot's gripper

3 3D Printing of the Prototype

From an (stl.) format of each robot's part presented in Sect. 2, the manufacturing design of the 3D printing process is done using ideaMaker software. It is a slicing software used for 3D printing. It is developed by Raise3D, a company that provides 3D printing solutions for a variety of industries (*IdeaMaker | Raise3D*, n.d.).

The ideaMaker software allows users to prepare their 3D models for printing by generating the G-code, which consists of a set of instructions that controls the 3D printer's movements and extrusion of filament.

The used RAISE3D N2 3D printer, available at ISSAT-SO, is designed for high-quality printing of complex objects and has a build volume of 305 mm × 305 mm × 305 mm. It is compatible with various filaments such as PLA, ABS, and PETG, and comes with the ideaMaker slicing software for customization of print settings and G-code generation. These features make it suitable for a range of applications, from prototyping to small-scale production.

The robotic arm in this project is fabricated using PLA filament, which is a popular material for 3D printing due to its low cost, ease of use, and good mechanical properties. PLA is also environmentally friendly due to its biodegradability and renewable resources. Its low melting point and ease of printing at lower temperatures compared to other filaments reduce the risk of warping and allow for high-quality prints, making it a suitable material for the intended robotic arm. The following table represents the principal parameters as set in the ideaMaker software (Fig. 6).

Fig. 6. RAISE3D N2 3D printer

According to the parameters in Table 1, Figs. 7, 8 and 9 provide a slicing insight of the base parts, the parallelogram linkage parts, and the gripper parts, respectively. The additive manufacturing process of this robot involved a computer aided manufacturing phase, which requires a specific tuning of printing parameters. By engaging in the additive manufacturing process, students are exposed to a computer-aided manufacturing phase where they need to fine-tune the printing parameters. This process not only

Table 1. The used 3D printing parameters.

Settings	Parameter	Value/Type
Layer	Layer heigh	0.3 mm
	Shell	2 mm
	Maximum Percentage of Shell Overlaps	50%
Extruder	Line width	0.4 mm
	Retraction speed	20 mm/s
	Quantity of the retracted material	1 mm
Infilling	Filling density	60%
	Outline overlap	15%
	Filling flow rate	100%
	Infill pattern	Grid
Full infilling	Number of bottom solid layers	3
	Number of upper solid layers	5
	Infilling layer of the low solid part	100%
	Infilling rate of the high solid part	100%
Support	Type of infilling	Lines
	Infilling rate	15%
Adhesion aid	Raft offset	5 mm
	Spacing between the raft and the model	0.3 mm
	Type of raft lines	Lines

provides hands-on experience but also requires students to make practical decisions and adjustments to achieve the desired results.

Assembling the 3D printed parts of the robot, allows students to develop essential skills in mechanical assembly. They gain a deeper understanding of how the different parts interact and work together to create a functioning robot. The assembly of the final 3D printed robot is presented in Fig. 10.

This 3D printed robot affords students the opportunity to learn basic skills and tackle problem-solving challenges in robot development from a practical approach. The 3D printed model serves as a valuable educational tool for teaching students about the fundamentals of forward and inverse kinematics in robot manipulators. In order to accomplish this, students were tasked with developing a graphical user interface (GUI) using the Matlab App Designer tool. The GUI provides a user-friendly control interface for the robot, enabling a direct control of the robot's joints and position control of the robot's end-effector. Figure 11 shows an example of a GUI's design view developed by students using Matlab App Designer.

The code developed for this GUI includes a kinematic model of the robot, as well as programming instructions for an Arduino board. These instructions enable control of

Fig. 7. Slicing insight of the base parts

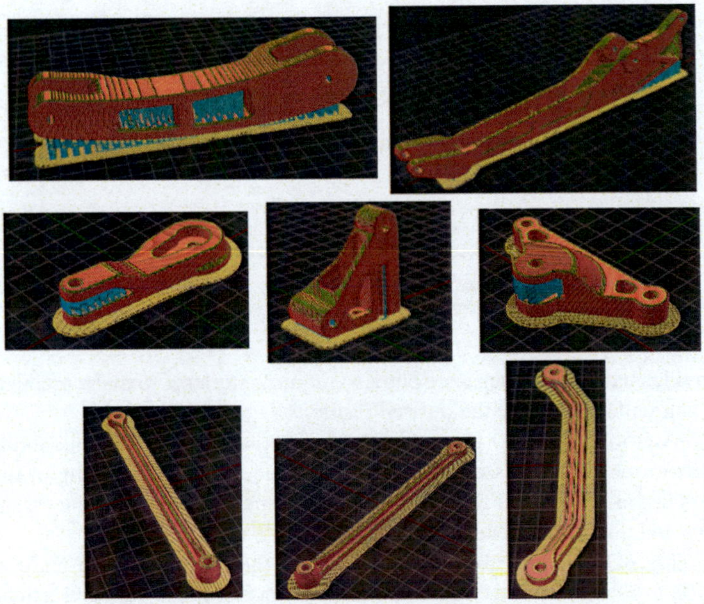

Fig. 8. Slicing insight of the parallelogram linkage parts

the four servomotors attached to each link of the robot, including the gripper's servomotor. The code considers user input values, allowing the robot to execute the desired movements based on the specified parameters. Figure 12 Presents a block digram of the developed GUI that summarize its methodology.

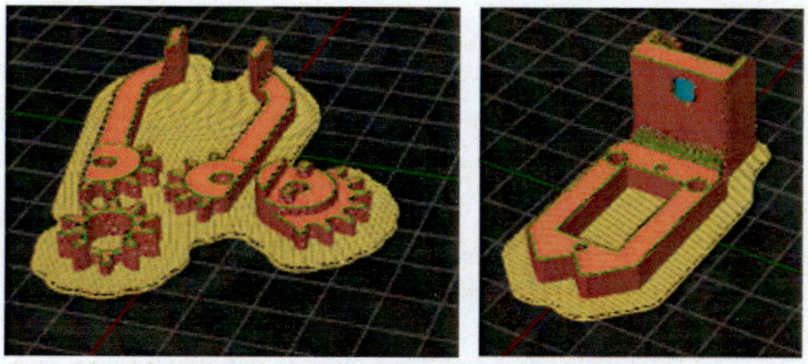

Fig. 9. Slicing insight of the gripper parts

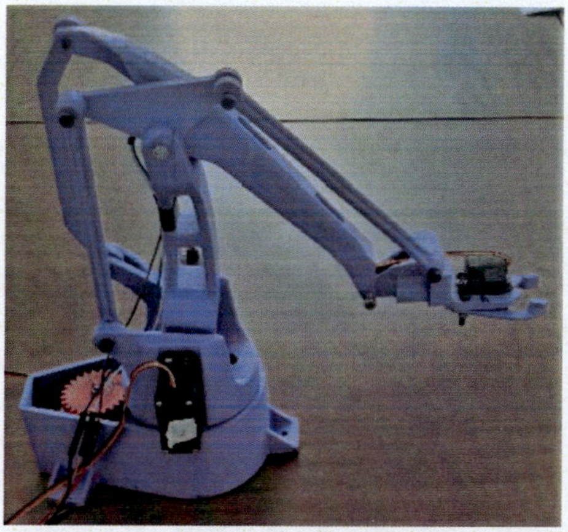

Fig. 10. 3D printed robot

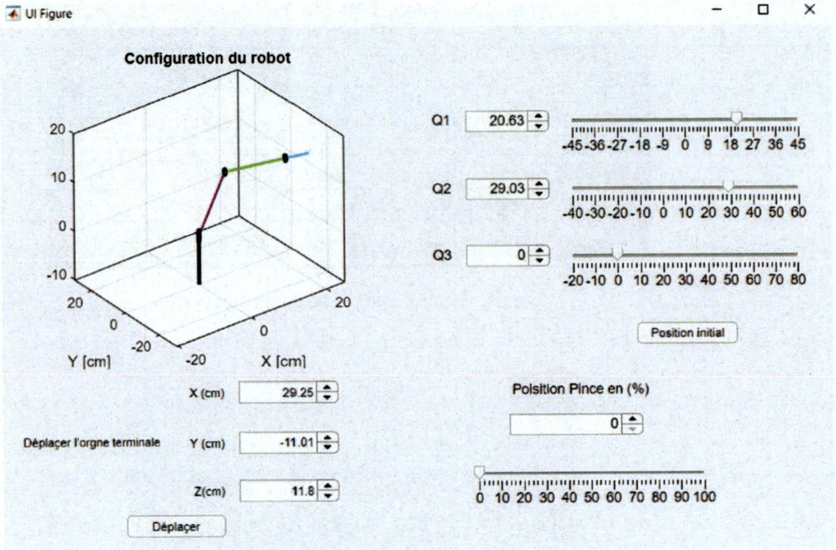

Fig. 11. GUI to control the robot movements

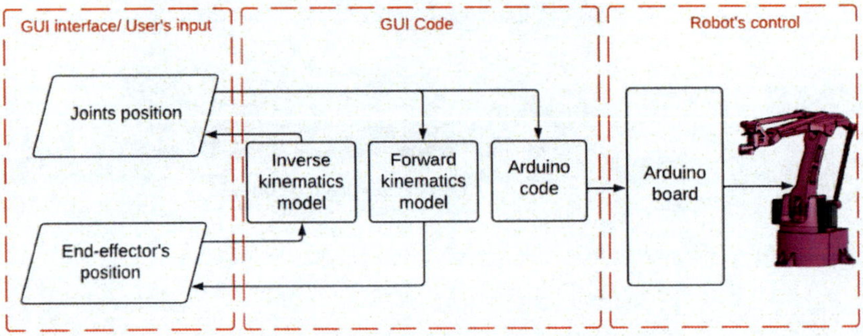

Fig. 12. Block diagram of developed GUI

4 Conclusion

In conclusion, this paper presents the 3D printing process of an open-source robot model that is proposed as an educational robot for ISSAT-So. Using 3D the FDM additive manufacturing technique, we were able to fabricate a robotic arm an open source robotic arm mimics the technical qualities and operating characteristics of an industrial robot, from a kinematic point of view. The fabricated robotic arm represents an affordable and accessible solution for educational purposes in the field of robotics, allowing students to gain hands-on experience in robot design, assembly, and programming. The integration of a graphical user interface (GUI) and the utilization of an Arduino board further enhance the learning experience, enabling students to explore embedded systems and gain programming skills. This work demonstrates the potential of 3D printing and additive manufacturing in the development of educational robot prototypes, and its role in advancing the field of robotics education. Further development in 3D printed educational robotics might include exploring dynamic capabilities in fabricated robot prototypes. By incorporating sensors and actuators, students can delve into the study of robot dynamics, gaining a comprehensive understanding of concepts like motion planning, trajectory generation, and feedback control. This enables a deeper exploration of the interaction between the robot and its environment.

References

Alimisis, D.: Educational robotics: Open questions and new challenges. Themes Sci. Technol. Educ. **6**(1), Article 1 (2013)

Altin, H., Pedaste, M.: Learning approaches to applying robotics in science education. J. Balt. Sci. Educ. **12**(3), 365–377 (2013)

Chavdarov, I., Nikolov, V., Naydenov, B., Boiadjiev, G.: Design and control of an educational redundant 3D printed robot. In: 2019 International Conference on Software, Telecommunications and Computer Networks (SoftCOM), 1–6 (2019). https://doi.org/10.23919/SOFTCOM.2019.8903825

Franciscone, C.: EEZYbotARM MK2 by daGHIZmo—Thingiverse (2016). https://www.thingiverse.com/thing:1454048

Gonzalez-Gomez, J., Valero-Gomez, A., Prieto-Moreno, A., Abderrahim, M.: A new open source 3D-printable mobile robotic platform for education. In: Rückert, U., Joaquin, S., Felix, W. (eds.) Advances in Autonomous Mini Robots, pp. 49–62. Springer, Cham (2012). https://doi.org/10.1007/978-3-642-27482-4_8

IdeaMaker | Raise3D. (n.d.). Raise3D: Reliable, Industrial Grade 3D Printer. Retrieved May 31, 2023, from https://www.raise3d.com/ideamaker/

IRB 460 | ABB Robotics. (n.d.). Robotics. https://new.abb.com/products/robotics/robots/articulated-robots/irb-460. Accessed 17 Mar 2023

Rosheim, M.E.: Robot Evolution: The Development of Anthrobotics, 1st edn. Wiley-Interscience (1994)

Telegenov, K., Tlegenov, Y., Shintemirov, A.: A low-cost open-source 3-d-printed three-finger gripper platform for research and educational purposes. IEEE Access **3**, 638–647 (2015). https://doi.org/10.1109/ACCESS.2015.2433937

Thrun, S., Sukhatme, G.S., Schaal, S. (eds.): Robotics: Science and Systems I (1st edition). The MIT Press (2005)

Topological Optimization of Ti-6Al-4V Knee Prostheses Through Lattice Structures for Enhanced Mechanical Performance

M. Frija[1,3](✉), A. Toumi[1,3], and M. Khodja[2,4]

[1] Institute of Applied Sciences and Technology of Sousse (ISSATSo), University of Sousse, Sousse, Tunisia
mounir.frija@issatso.u-sousse.tn
[2] Faculty of Engineering, Built Environment and Information Technology, University of Pretoria, Pretoria, South Africa
[3] Unit of Mechanical Production Engineering and Materials (UGPMM), UR17ES43, B.P. 1173, 3038 Sfax, Tunisia
[4] Materials Science and Metallurgical Engineering, University of Pretoria, Pretoria, South Africa

Abstract. In this study, we performed a topological optimization on a knee prosthesis made of Ti-6Al-4V titanium super alloy using metallic additive manufacturing. The optimization was done using nTopology software, wherein lattice structures were incorporated within the femoral component to reduce its weight while maintaining adequate mechanical strength. A comparative analysis of the prosthesis' mechanical strength with and without optimization was conducted. Our simulation results, obtained through ANSYS, reveal the advantages of topological optimization in the medical industry.

Keywords: Additive manufacturing · Topology Optimization · Lattices structure · Knee prosthesis · nTopology · Ti-6Al-4V

1 Introduction

The literature presents various approaches to design parts for additive manufacturing, including topological or parametric optimizations, design rules derived from process characterization, and specific tools [1–3]. Topological optimization is a popular method used for additively manufactured parts, aiming to optimally distribute material within a defined design space while respecting mechanical constraints and user requirements (Fig. 1). Topology-optimized design differs from traditional design in that it can achieve functional specifications in a single iteration [4]. Studies in orthopedic biomechanics have shown that topological optimization of prosthesis design can improve implant lifespan [5, 6].

This paper focuses on topological optimization for additive manufacturing, with an application on a knee prosthesis using the software **"nTopology".** As the simulated part was too heavy, we optimized the calculation time by considering the part as symmetrical.

T. Mabrouki et al. (Eds.): AIAM 2023, LNME, pp. 70–79, 2024.
https://doi.org/10.1007/978-3-031-47784-3_9

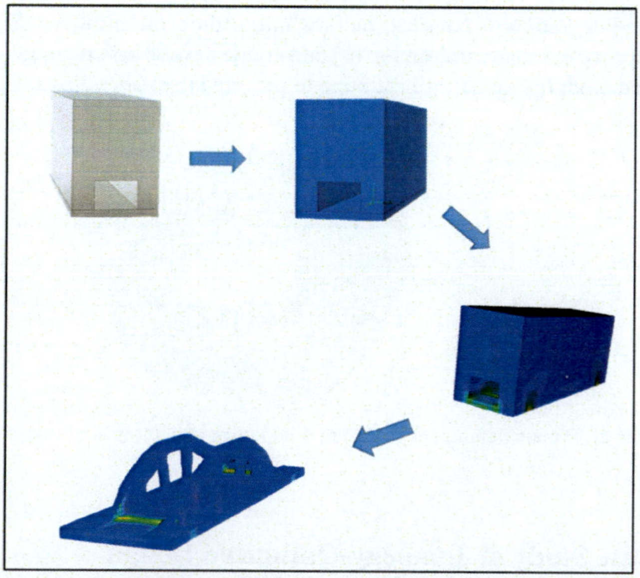

Fig. 1. Topology optimization

2 Topology Optimization for Additive Manufacturing

Topology Optimization (TO) has experienced significant growth in both academic research and industrial applications, thanks in large part to additive manufacturing (AM). AM enables the production of highly complex parts with intricate shapes that are difficult to achieve through conventional machining methods. This makes AM the preferred method for obtaining topology-optimized parts. The use of topological optimization tools in combination with AM reduces material and manufacturing time, improves the design process, and leads to high-performance products with minimized compliance and mass [7]. Additionally, this innovative approach can suggest unconventional shapes that designers may not have considered. The combination of TO with AM leads to light weighting, improved structural performance, and faster design-to-manufacturing cycles, making it particularly relevant for industries such as automotive, aerospace, and medical, where weight is a critical factor.

Additive manufacturing combined with topology optimization has enabled the production of personalized lightweight prosthetic limbs, cranial implants, and orthopedic implants such as hips and knees [8]. This process allows for the successful construction of structures with porous infill, leading to the development of overhang-free infill topology optimization methods. Lattice structures, which are repetitive unit cells composed of strips or lattices with various patterns such as octagonal, honeycomb or random patterns, are the most commonly used infill designs (Fig. 2). These structures are bio-inspired and offer solutions for weight, energy, and advanced manufacturing time reduction.

The integration of lattice structures into orthopedic implants shows promise, as it can improve osseointegration [9, 10] and create desirable structural properties such as a matched modulus to that of host bone to prevent stress shielding [11].

Fig. 2. Porous titanium cage design with lattice structures to promote bony in-growth

3 Case Study of Topology-Optimized Design

3.1 The General Workflow

The optimization process for the knee implant consists of three main steps. Firstly, a static analysis is performed on the implant using ANSYS software. Following this, the interior of the femoral part of the implant is filled with lattice structures using the "nTopology" software to carry out topological optimization (Fig. 3). Finally, the optimized part undergoes validation through another static analysis.

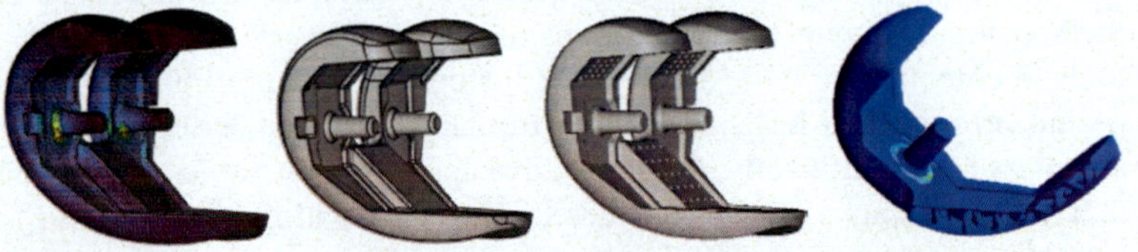

Fig. 3. General workflow

3.2 The Chosen Software for Optimisation

The software chosen for topological optimization is nTopology, which is a cutting-edge engineering software, aimed at developing a better product development system based on data and physics. This software is specifically designed for topological optimization, offering various methods to lighten parts, such as different types of lattice structures and perforation. It is also unique in integrating design, simulation, and manufacturing tools within a single environment, making it an ideal choice for advanced manufacturing, including additive manufacturing.

3.3 Identification of the Implant: Knee Prosthesis

The knee prosthesis is generally composed by three components namely; the femoral component, the tibial tray and the tibial insert [12]. Figure 4 shows an illustration of an integrated knee implant.

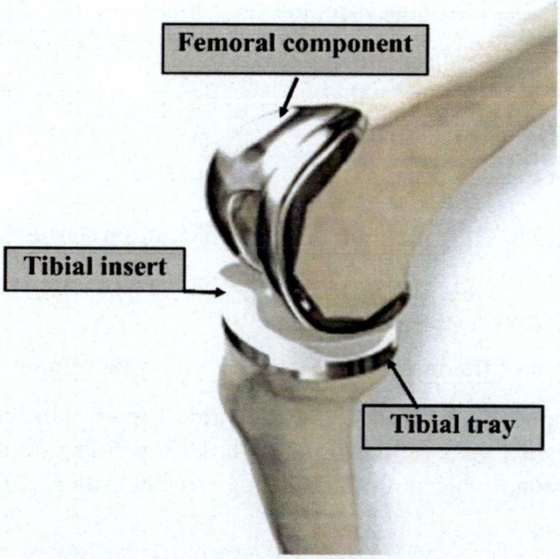

Fig. 4. Elements of a knee joint prosthesis

The geometric CAD model designed by CATIA V5. As the part is symmetrical, we decided to apply the static analysis before and after optimization only for half of the part (Fig. 5).

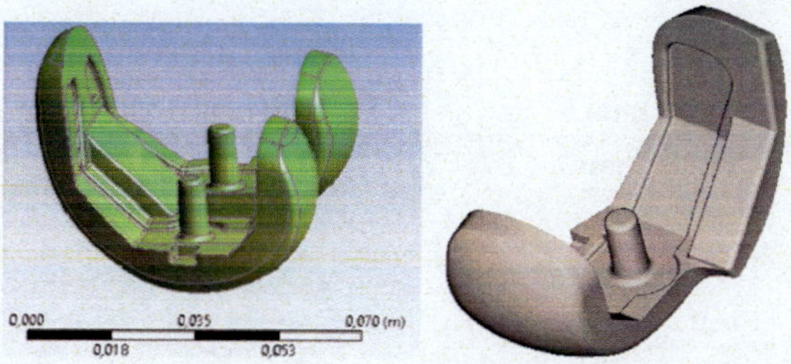

Fig. 5. From Initial CAD model of the femoral component to a reduced model "Half of the part"

3.4 Material Identification

The choice of material for this study is the Ti-6Al-4V titanium alloy due to its low density and low modulus of elasticity, which makes it more compatible with bone tissue. Table 1 presents the various parameters of the titanium alloy used in the research.

Table 1. Properties of the Ti-6Al-4 V titanium alloy

Properties	Value
Density (kg.m^{-3})	4429
Poisson's ratio	0,3387
Young's modulus (MPa)	1,112 E^5
Modulus of elasticity (MPa)	1,149 E^5
Shear Modulus (MPa)	4,1533 E^4
Tensile strength (MPa)	8,547 E^2

3.5 Topological Optimization of Femoral Component by Lattice Structures

The topological optimization by lattice structures on the nTopology software is performed as shown below:

a) **Conversion of the imported part into an implicit body**

Upon importing the CAD part, it is transformed into the implicit format, which provides a lightweight yet powerful method for defining, representing, and modifying three-dimensional objects through a single mathematical function that represents the solid.

b) **b- Internal volume lightening by lattice structures**

The "Shell and Volume lattice" block in nTopology simplifies the application of complex structures, making it easy and practical. In our specific case, this block involves breaking down an implicit body and filling it with volume lattices (Fig. 6) (Table 2).

Table 2. Parameter for lightening the implant

Body	(Implicit Body)		
Type of lattice	body-centered cubic		
Cell size (mm)	10	10	10
Rotation	0	0	0
Lattice thickness (mm)	1		
Shell thickness (mm)	1		
Mixing radius (mm)	0.5		

After this operation, the new weight of the piece will be equal to 62 g. These lattices allowed us to lighten it by more than half of material.

c- Creation of powder discharge events

To achieve significant weight reduction, it is crucial to remove any unfused powder from the interior of the part after its laser powder bed fusion 3D printing. Otherwise,

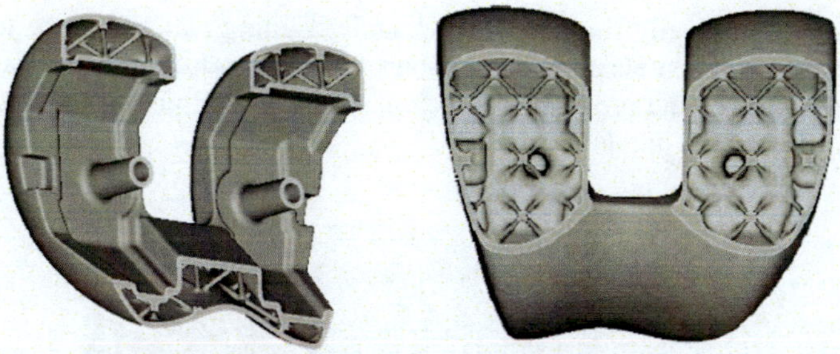

Fig. 6. Sectional view after internal lightening

the part would remain filled with excess powder. To address this issue, we opted to create perforations on the part's surfaces, as shown in Fig. 7, to guide the removal of the remaining material (Table 3).

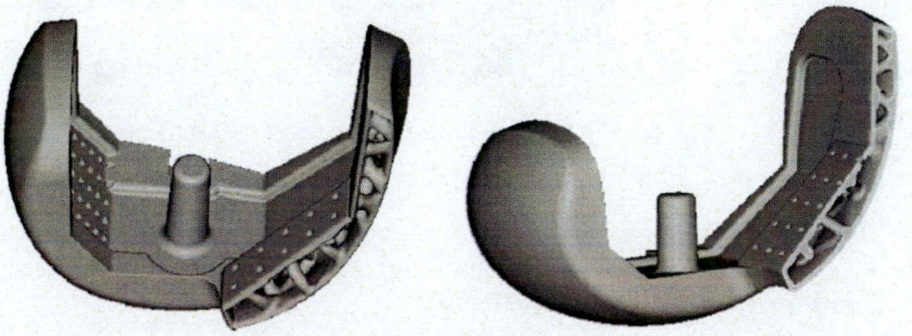

Fig. 7. Half of the part to be exported for analysis

3.6 Mesh of the Part

Figure 8 shows the part after meshing.

Table 3. The part mesh specifications

Mesh Type	triangular
Number of nodes	285864
Number of elements	1282327

Meshing is the process of dividing a 3D model into smaller and simpler parts, called mesh elements, to prepare it for analysis or simulation. In the context of a knee prosthesis, meshing involves creating a finite element mesh of the prosthetic component, which is then used to simulate its behavior under various loading conditions. The finite element mesh is created by dividing the 3D CAD model of the prosthetic component into small, geometrically simple elements, and assigning material properties and boundary conditions to each element. This allows for a more detailed analysis of the mechanical behavior of the prosthesis, which can help in optimizing its design for improved performance and durability

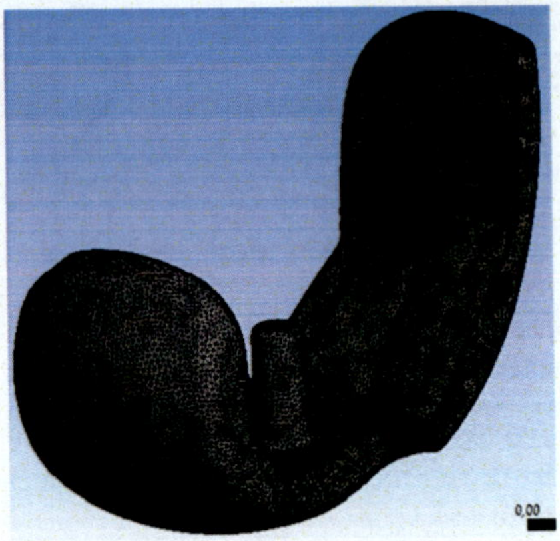

Fig. 8. CAD model mesh of the femoral half component

3.7 Load and Boundary Conditions

The knee joint is a multifaceted structure in the human body that undergoes high loads during various physical activities like walking, running, sitting, and rotational movements. According to studies, the maximum compressive forces on the knee joint during daily activities range **from 4 to 4.5 times** body weight [8]. In the figure, the blue-colored cylindrical shapes are affixed to the upper part of the knee, while a load of 2200 N along the z-axis (weight = 55 kg) is applied on the surface depicted in red (Fig. 9). The load during the walking cycle according to the bibliographical references is estimated 4 times that of the static load.

4 Results and Discussion

Following a comparative study of the results before and after optimization, we can summarize the results obtained by the static analysis in Table 4.

(See Fig. 10).

At first, the femoral component of the knee prosthesis was subjected to the same loads and boundary conditions as shown in Fig. 9, where the red and blue colors represent them. The stress distribution in Fig. 11 indicates that the areas affected in the part with lattice structures have reduced, although the values have increased as per Table 4.

The primary objective of optimizing this femoral part is to reduce material usage while maintaining the Von Mises stress values below the yield strength of Ti6Al4V

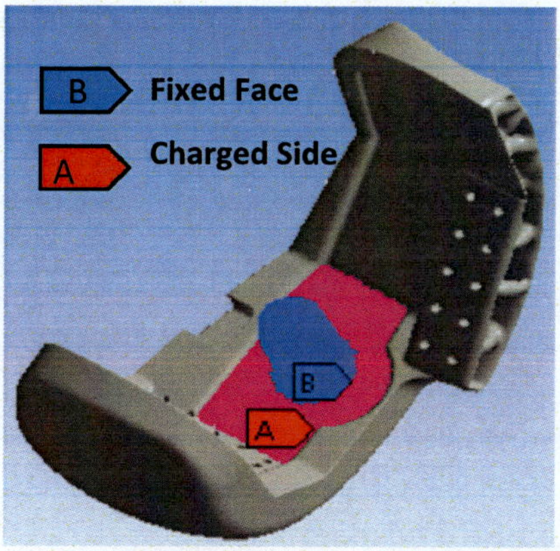

Fig. 9. Application of loads and boundary conditions

Table 4. Comparison of results before and after optimization

		Part before optimization	Part after optimization
Von Mises stress (MPa)	Max	**75**	**448**
	Min	**0**	**0**
Distortion	Max	**9.e^{-10}**	**4.e^{-9}**
	Min	**0**	**0**
Max total displacement (m)		**9.e^{-9}**	**5.e^{-8}**
Total weight (g)		**154**	**78**

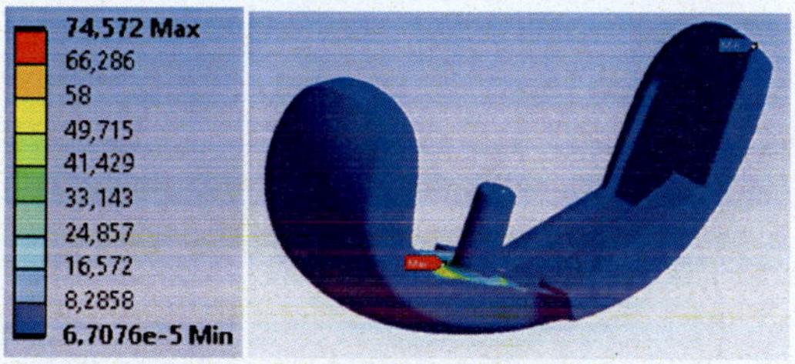

Fig. 10. Stress distribution inside the part without topological optimization

titanium, which is 863 MPa, ensuring that the part remains in the safe elastic zone. The optimized femoral part has achieved almost half of its initial weight, with a maximum value of Von Mises far below the elastic limit of the material, indicating that the part is safe in the elastic zone. Therefore, the new optimized part is deemed effective .

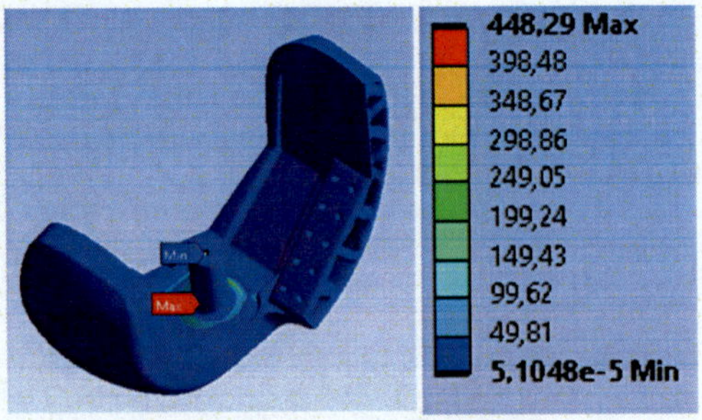

Fig. 11. Stress distribution inside the part after topological optimization

5 Conclusion and Perspectives

The research project presented in this paper utilizes a topological optimization approach to reduce the weight of components by incorporating lattice structures within their volume. To this end, we employed the nTopology software due to its flexibility and effectiveness in such applications. To validate our optimization, we carried out numerical simulations using the Ansys finite element simulation software. Additionally, we suggest that future research in this area should explore the use of surface lattice structures that mimic the Young's modulus of bone tissue to enhance osseointegration with the implant surface. This approach could potentially alter the surface morphology of implants in direct contact with the bone.

Acknowledgements. The authors express their gratitude for the financial assistance provided by the CSIR National Laser Centre in Pretoria, South Africa [CSIR-NLC Reference: HLHA23X task ALC-R008].

I express my heartfelt appreciation to the nTopology Team for granting me the educational license that was instrumental in the successful completion of this research paper.

References

1. CETIM: Fabrication additive métallique – Les fondamentaux , Centre Technique Des Industries Mécaniques (2019)
2. Dutta, B., Froes, F.H.: Additive manufacturing of titanium alloys. state of the art, challenges and opportunities. Elsevier Inc., (2016). ISBN: 978-0-12-804782-8
3. Lin, C.: Biomedicine. In: Tech Editor, chap 7 : Additive Manufacturing Solutions for Improved Medical Implants, pp. 147–180 (2015). ISBN-13 : 978-9535103523
4. Gebisa, A.W., Lemu, H.G.: A case study on topology optimized design for additive manufacturing. In: IOP Conference Series: Materials Science and Engineering, vol. 276, pp. 012026 (2017). https://doi.org/10.1088/1757-899x/276/1/012026
5. Mohd, J., Abid, H.: Additive manufacturing applications in medical cases: a literature based review, Alexandria J. Med. **54**(4) (2018). https://doi.org/10.1016/j.ajme.2017.09.003
6. Azman, A.H., Vignat, F., Villeneuve, F., Nguyen, D.S.: Creation of lattice structures with skeleton model for additive manufacturing. Int. J. Interact. Des. Manuf. (IJIDeM) **15**(4), 381–396 (2021). https://doi.org/10.1007/s12008-021-00767-z

7. Louvigny, Y., Nzisabira, J., Meunier, B. Duysinx P.: L'optimisation topologique et la fabrication additive, amélioration de la chaine de conception. In: 14e Colloque National AIP-Priméca , La Plagne, vol. 73, pp. 31 Mars au 02 avril (2015)
8. Pavan, V., Kumar, M., Bhanulatha, I.: Design and structural analysis of knee implants using different materials. Int. J. Adv. Sci. Res. Eng. Trends **5**(7), 2456–0774 (2020). ISSN (Online)
9. Ghilini, F., Fagali, N., Pissinis, D.E., Benítez, G., Schilardi, P.L.: Multifunctional titanium surfaces for orthopedic implants: antimicrobial activity and enhanced osseointegration. ACS Appl. Bio Mater. **4**(8), 6451–6461 (2021). https://doi.org/10.1021/acsabm.1c00613
10. Liu, Y., Rath, B., Tingart, M., Eschweiler, J.: Role of implants surface modification on osseointegration: a systematic review. J. Biomed. Mater. Res. Part A (2019). https://doi.org/10.1002/jbm.a.36829
11. Park, P.J., Lehman, R.A.: Optimizing the spinal interbody implant: current advances in material modification and surface treatment technologies. Curr. Rev. Musculoskelet. Med. (2020). https://doi.org/10.1007/s12178-020-09673-5
12. Hafezalkotob, A., Hafezalkotob, A.: Comprehensive MULTIMOORA method with target-based attributes and integrated significant coefficients for materials selection in biomedical applications. Mater. Des. **87**, 949–959 (2015). https://doi.org/10.1016/j.matdes.2015.08

Manufacturing of Cranial Implant Using SPIF and 3D Printing

Sofiene Marzouki[1](✉), Slim Bouzidi[2], and Atef Boulila[2]

[1] Higher Institute of Technology Study of Jendouba, Campus Universitaire, 8189 Jendouba du Nord, Tunisia
Sofiene_marzouki@hotmail.fr

[2] National Institute of Applied Sciences and Technology, University of Carthage, Centre Urbain Nord, Tunisia

Abstract. The single-point incremental forming (SPIF) and fused deposition modelling 3D printing (FDM) are used in the manufacture of parts for small series and prototypes in various sectors such as aeronautics, biomedical field, and art pieces. In this sense, recent published results have revealed an increasing interest in shaping components for biomedical uses. This work aims to study the geometric precision of the two methods during the development of a cranial implant and to compare the fidelity of the parts to be developed to the CAD part

Keyword: SPIF · FDM · biomedical · Cranial Implant

1 Introduction

Custom-made cranial implants are medical devices that differ in size and configuration (directly dependent on the size and location of bone loss) and are used during cranioplasty to correct bone loss from the skull bones.

The main objectives of cranioplasty are to prevent the risk of injury and infection of the unprotected brain (Rotaru 2012), to correct malformations of the skull bones and to restore its mechanical and aesthetic aspects.

Skull bone recuperation is a delicate process that poses a challenge to the surgical staff (Argenta 1996). Due to design complexity and material Compatibility, the primary enterprises are implant design and material selection.

Despite advances in 3d printing and the spif process, there are many difficulties encountered; including geometric accuracy, production time and finished product condition (Bouzidi 2022).

T. Mabrouki et al. (Eds.): AIAM 2023, LNME, pp. 80–86, 2024.
https://doi.org/10.1007/978-3-031-47784-3_10

2 Methodology

The production of the implant is divided into several steps (Fig. 1).

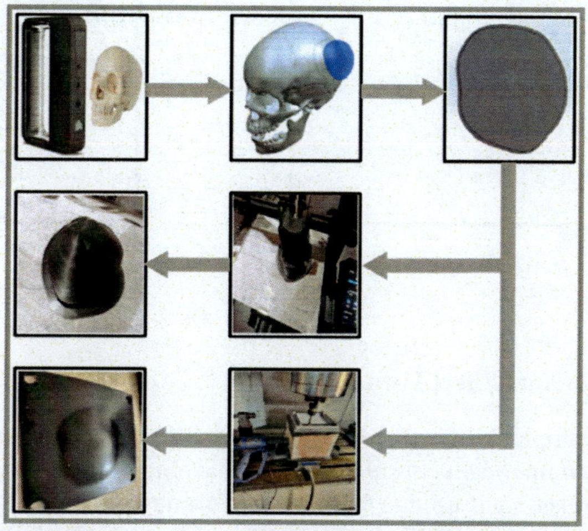

Fig. 1. 3D printing and SPIF cranial implant manifacturing process

2.1 Scan, CAD, CAM and 3d Processing:

3d scan of the skull with a scanner sense 2 (Fig. 2) and localization of the study area using CAD software (Fig. 3) (Skjoet 2007), then we generate the trajectory that will be used in the manufacture of the part by SPIF method (Fig. 4) (Boulila 2018), we generate also an STL file that will be used for 3d printing (Fig. 5).

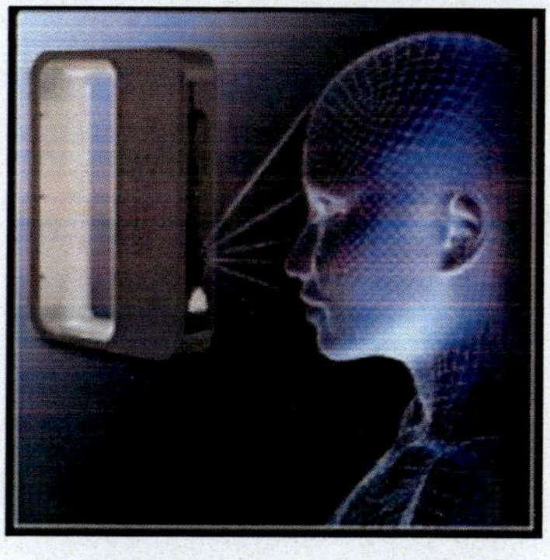

Fig. 2. 3d scan of skull

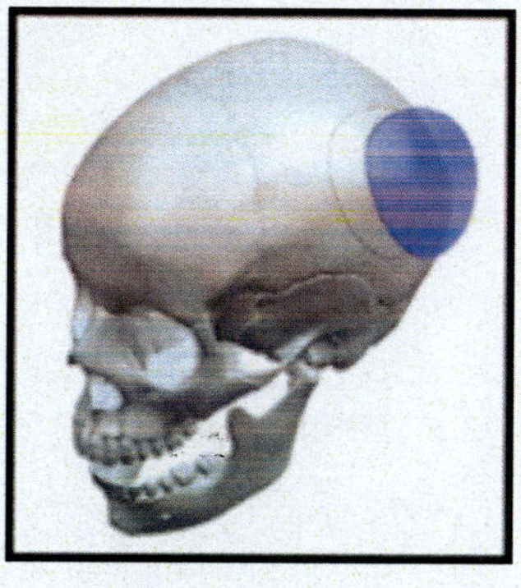

Fig. 3. Study area

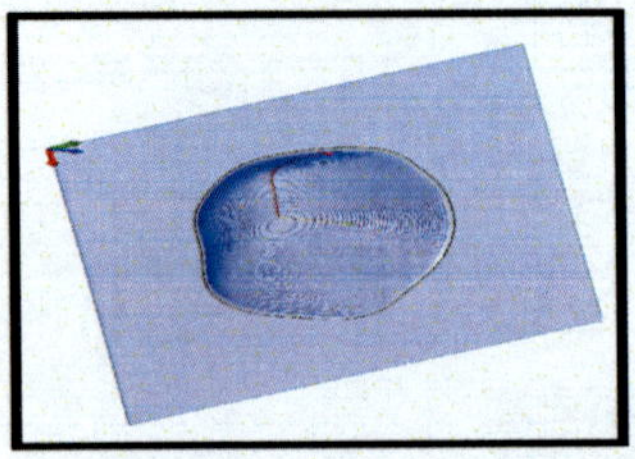

Fig. 4. Tool path

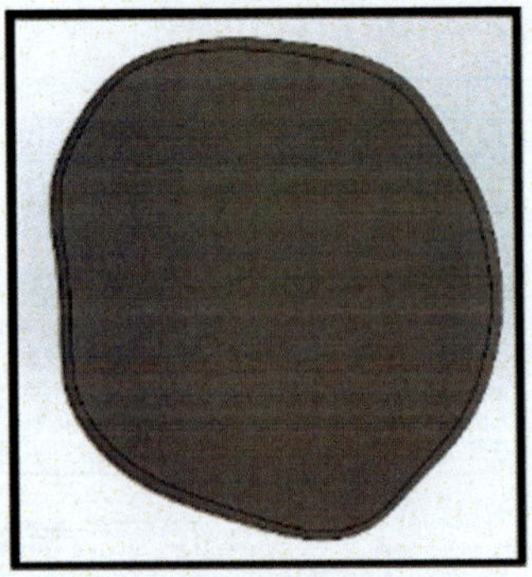

Fig. 5. STL file

2.2 Incremental Sheet Forming (SPIF)

The cranial implant is obtained by incremental forming of a 3 mm thick sheet of HDPE on a 3-axis numerical controlled milling machine Spinner MVC850 (Fig. 6), a device with hot air heating is used to facilitate the forming process (Fig. 7), the mechanical and thermal propriety of HDPE sheet (Fig. 8) are given in (Table 1) (Khelif 2008).

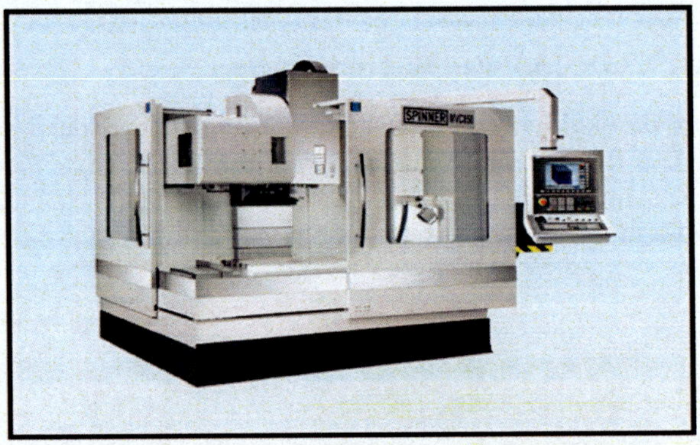

Fig. 6. CNC milling machine Spinner MVC850

2.3 3D Printing (FDM)

Another part is obtained by 3d printing from a CAD model "STL file" (Fig. 4) using a 3D Printer "Creality Ender-3 v2" (Figs. 9 and 10) (Bakkiyaraj 2021).

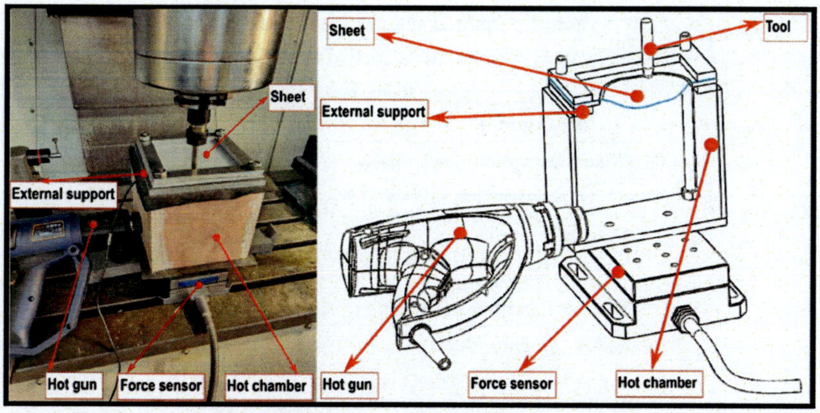

Fig. 7. Heat forming device

Fig. 8. Finished SPIF Part

2.4 3D Scan of the Finished SPIF and Printed Part.

The finished parts are scanned by a 3d scanner (Fig. 11) then the files are processed on a mesh processing software MeshLab

3 Results

To superimpose the finished parts obtained by SPIF and 3D printing with the CAD models, we used the MeshLab software ICP (Fig. 12) (Cignoni 2011), then we cut out the intersection by planes along the three directions x, y and z and we compare the variation in distance between the CAD model and the finished part (Figs. 13–18).

We note that the geometric error during the manufacture of the cranial implant by 3D printing is less than that of the part obtained by incremental forming. The 3d printing

Table 1. Thermal and mechanical properties of HDPE

Property	Value
Density (kg/m^3)	950
Water absorption (24 h basis)	<0.01%
Thermal expansion (10^6 K^{-1})	100–200
Poisson's ration	0.46
Friction factor	0.29
Rockwell hardness (shore)	D 60–73
Elastic modulus (Gpa)	0.5–1.2
Impact resistance, IZOD (j/m)	20–210
Yeld stress (Mpa)	15–40

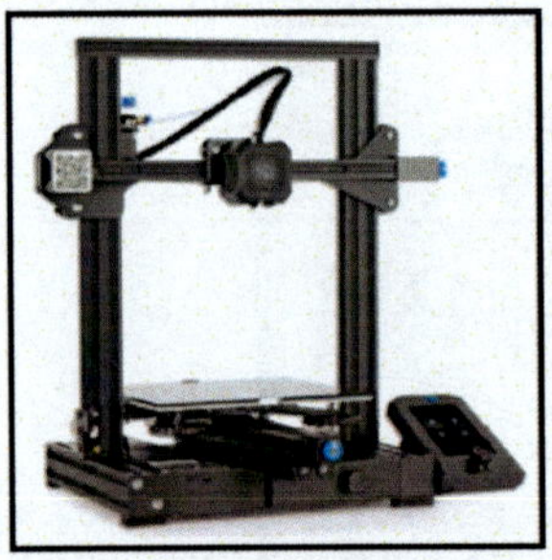

Fig. 9. Creality Ender-3 v2 3D Printer

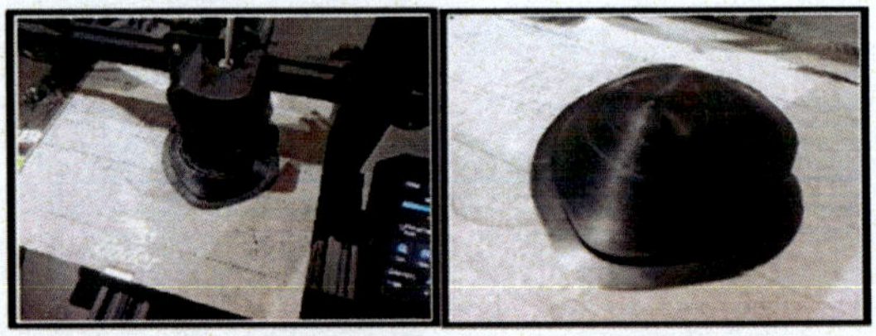

Fig. 10. Finished printed part

Fig. 11. 3D Scan of the finished part

take more time then SPIF operation more than 8 times slower then SPIF and also the 3d printing surface is rougher then SPIF.

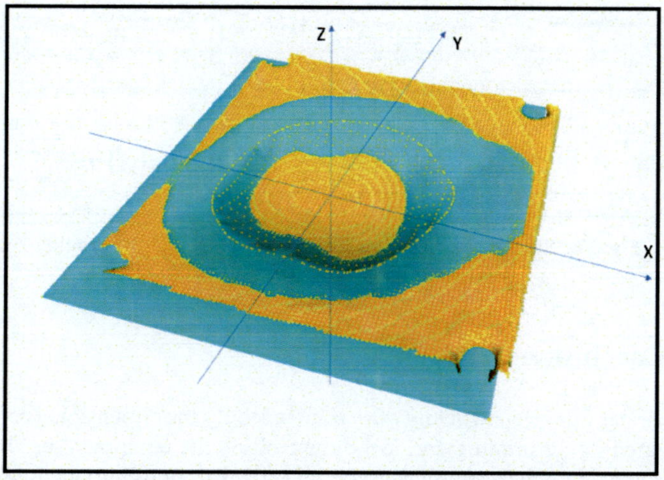

Fig. 12. Superimpose CAD and finished part

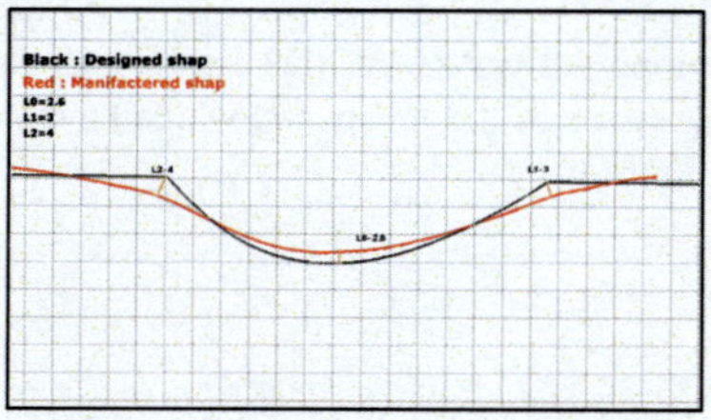

Fig. 13. SPIF X plane intersection

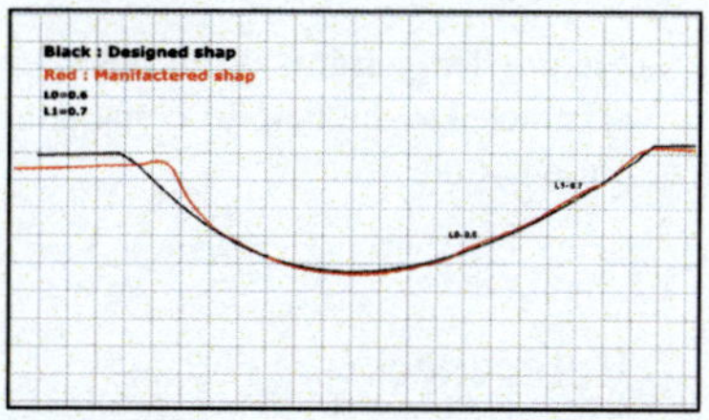

Fig. 14. 3D printing X plane intersection

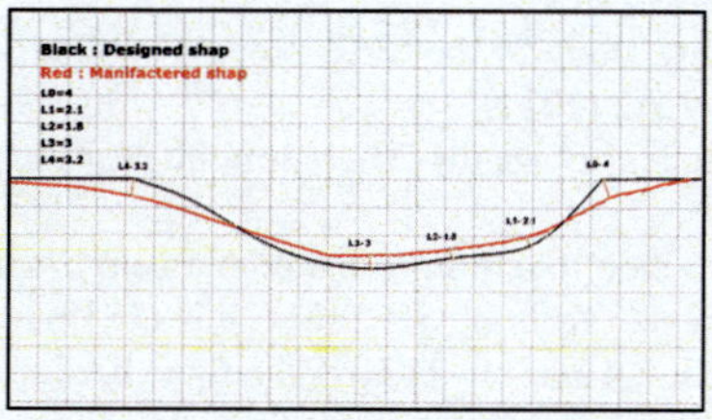

Fig. 15. SPIF Y plane intersection

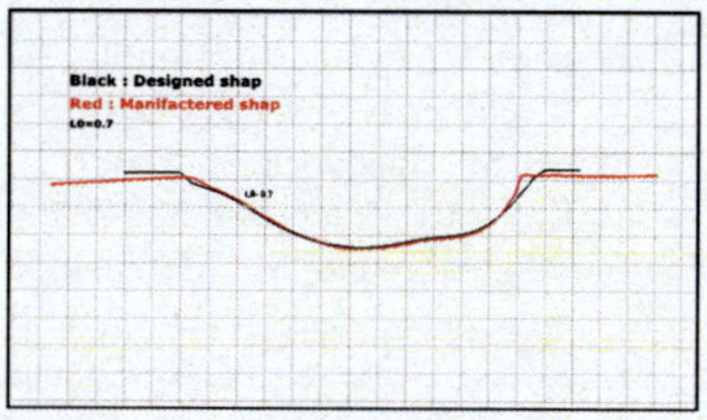

Fig. 16. 3D printing Y plane intersection

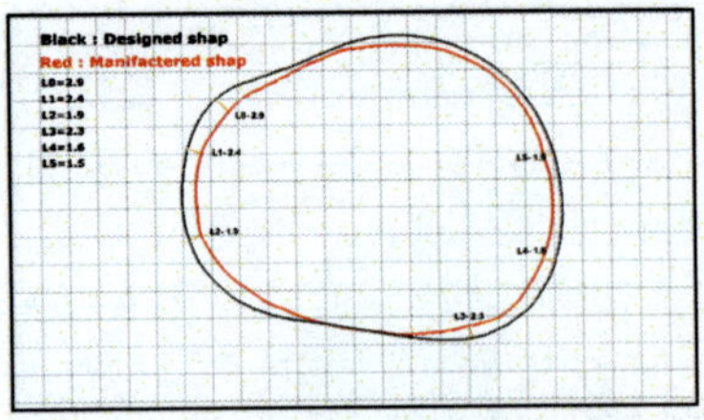

Fig. 17. SPIF Z plane intersection

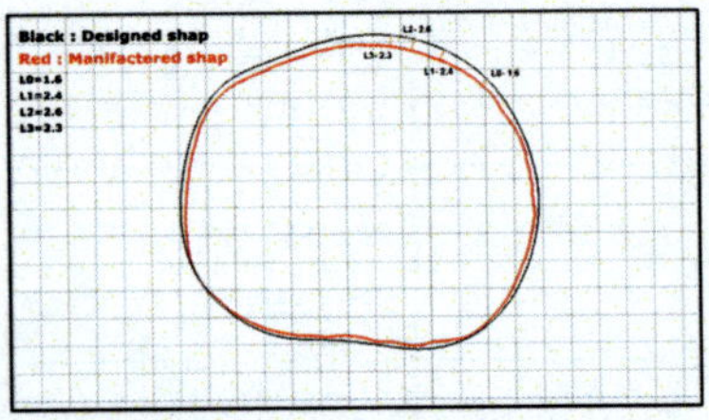

Fig. 18. 3D printing Z plane intersection

4 Conclusion

The strategy developed in this work is intended for the dimensional control of parts manufactured by 3d printing and ISF by exploiting the equipment used in the production line and without the need for new control and inspection equipment. A dimensional mismatch has been made between the design shape of a CAD model and the corresponding manufactured part.

Acknowledgements. This work is partially supported by AIAM'2023 The authors also gratefully acknowledge the helpful comments and suggestions of the reviewers, which have improved the presentation.

References

Rotaru, H., Florian, I., Park, Y., Balc, N.: Cranioplasty with custom-made implants: analyzing the cases of 10 patients. J. Oral Maxillofac. Surg. **70**, 169–176 (2012)

Argenta, L., David, L., Wilson, J., et al.: An increase in infant cranial deformity with supine sleeping position. J. Craniofac. Surg. **7**, 5–11 (1996)

Bouzidi, S., Ayadi, M., Boulila, A.: Feasibility study of the SPIF process applied to perforated sheet metals. Arab. J. Sci. Eng. **47**, 9225–9252 (2022)

Skjoet, M., Handcock, M., Bay, N.: Creating helical tool paths for single point incremental forming. Key Eng. Mater. **344**, 583–590 (2007)

Boulila, A., Ayadi, M., Marzouki, S.: Contribution to a biomedical component production using incremental sheet forming. Int. J. Adv. Manuf. Technol. **95**, 2821–2833 (2018)

Khelif, R., Chateauneuf, A., Chaoui, K.: Statistical analysis of HDPE fatigue lifetime. Meccanica **43**, 567–576 (2008)

Bakkiyaraj, M., Kavitha, G., Krishnan, G., Kumar, S.: Impact of Augmented Reality on learning Fused Deposition Modeling based 3D printing Augmented Reality for skill development. Mater. Today Proc. **43**, 2464–2471 (2021)

Cignoni, P, Ranzuglia, G, Callieri, M, et al.: MeshLab: an open-source mesh processing tool. In: Eurographics Italian Chapter Conference 2008, pp. 129–136 (2011)

New Horizons in Engineering Education: From Additive Manufacturing to Immersive Learning

Safa Bhar Layeb[1(✉)], Amel Jaoua[1], and Mikel Noomen[2]

[1] LR-OASIS, National Engineering School of Tunis, University of Tunis El Manar, Tunis, Tunisia
{safa.layeb,amel.jaoua}@enit.utm.tn

[2] National Engineering School of Tunis, University of Tunis El Manar, Tunis, Tunisia
mikel.noomen@enit.utm.tn

Abstract. Additive Manufacturing (AM) has now become a fully-functional production process, which is becoming more and more widespread in various industrial sectors. Besides, AM has the potential to enhance learning by providing engineering students with hands-on, experiential and interactive learning experience. In the same line, higher education's focus on immersive learning has gained significant traction, thanks to the rise of digital technologies. In this context, this work reports the project conducted in the learning factory developed at the Industry 4.0 laboratory of the National Engineering School of Tunis (ENIT). Current projects are presented and challenges are pointed out.

Keywords: Additive manufacturing · immersive learning · engineering education

1 Introduction

Additive Manufacturing (AM) is a transformative approach to industrial production that enables the creation of parts and systems (ISO ASTM 59900 2015). It is a broad term that refers to a variety of methods capable of manufacturing 3D objects by adding layers on top of one another (Butt, 2020). In fact, AM has been commonly used for rapid prototyping purposes in a variety of industries which is the process of rapidly creating and prototyping a model or a part of it. They consist of building an object from scratch or from a semi-finished part acting as substrate.

Due to the pervasive adoption of AM practices in Industry 4.0, it becomes crucial to provide education and training to young engineers to effectively utilize these technologies. In fact, nowadays, several high-ranked engineering schools have gone further than training their engineering students to AM by using it as a part of an Immersive Simulation-Based Learning (ISBL) methodology (Melo et al. 2022). This method combines the benefits of immersive simulated environments and problem based learning.

In this context, this paper reports the project conducted at the learning factory established in the Industry 4.0 Laboratory, called Lab 4.0, at the Department of Industrial Engineering of the National Engineering School of Tunis (ENIT). It presents the current

T. Mabrouki et al. (Eds.): AIAM 2023, LNME, pp. 87–91, 2024.
https://doi.org/10.1007/978-3-031-47784-3_11

situation of using AM within the Lab 4.0 and provides avenues of integrating ISBL to enhance engineering education at ENIT.

The reminder of the paper is as follows. Section 2 presents the investigated laboratory. Section 3 describes the intersection of additive manufacturing and immersive learning and Sect. 4 reports their impact on engineering education. Section 5 drives conclusion and provides avenues for future work.

2 Additive Manufacturing at Lab 4.0

This study is part of the development of the ENIT Industry 4.0 Laboratory. The laboratory was acquired for the NePRev (NExt Production REVolution) project at the National School of Engineers of Tunis, and it integrates technological functionalities of Industry 4.0. This Industry 4.0 Laboratory comprises a manufacturing system that includes Festo didactic MPS 403-1 which consists on the distribution joining and sorting stations. These stations produce parts that are carried by an autonomous mobile robot to the Assembly station. In this station parts are assembled in the adequate package. These packages are produced using AM. For that purpose, the Lab 4.0 contains two Creality Ender 5 plus 3D printers. These printers have been used to produce faces for the different package's faces. Figure 1 shows the 3D printers used by the engineering students at the laboratory.

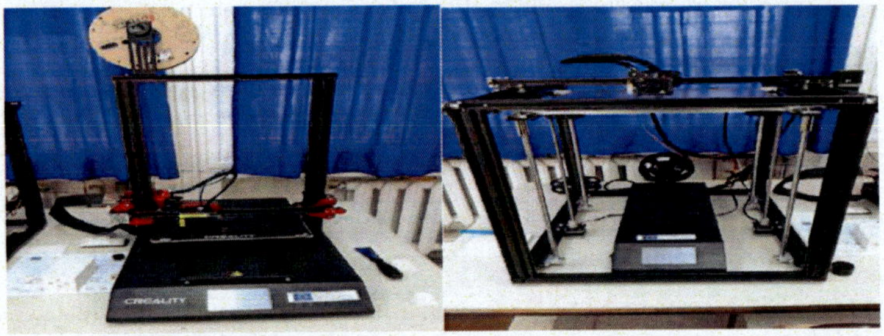

Fig. 1. Additive Manufacturing Machines used in ENIT Industry 4.0 Laboratory

In the Lab 4.0, these 3D printers have been used to create prototypes of sustainable reusable package, allowing students to experiment with and learn about sustainable design principles such as design for disassembly and design for adaptability in the settled learning factory. The process flow in this learning factory is displayed in Fig. 2.

Incorporating additive manufacturing education into curriculums is a crucial step in equipping young engineers with the skills and knowledge they need for green transition and Industry 4.0. It is worthy to mention that this laboratory is the site of several courses, such as operations management and digital twin for smart manufacturing courses using SIMIO as a simulation software. It also hosts several end-of-year engineering projects which contribution has been concretized by scientific publications (Jaoua et al. 2022; Jaoua et al. 2023).

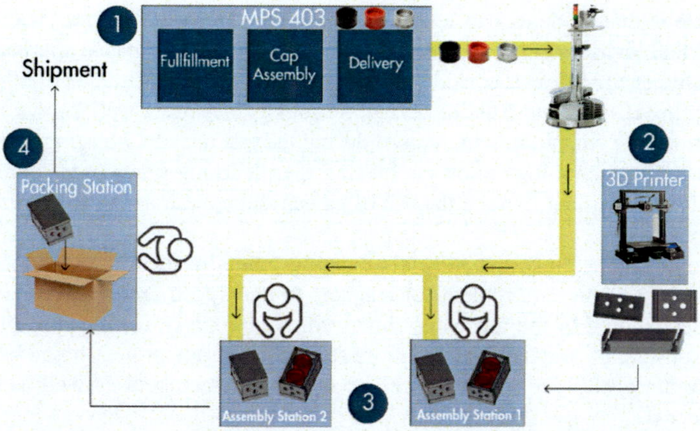

Fig. 2. Scheme of the learning factory at ENIT Industry 4.0 Laboratory

3 Toward Immersive Simulation-Based Learning

In the last decade, with the emergence of digital technologies, innovative teaching, based on immersive learning in higher education, is becoming prominently attractive. Implementation and the benefits of immersive learning in higher education have recently been depicted in (Marks and Thomas 2022). This learning approach combines virtual, augmented and mixed reality with simulation learning, which allows enhancing the learning experience by immersing students in simulated environments that mimic real-world situations. The goal is to provide students with more interactive learning by innovative, more naturalistic presentation of the information and reduction of learner's cognitive load. Besides, immersive experiences bring many advantages to learn in situations where being physically present would be too expensive, dangerous or impossible. Learners can interact with objects, and the Virtual Reality platform simulates how those objects would change in reaction. This way, allows learners to visualize concepts in a three dimensional space and manipulate them in a game-like environment. Students will have hands-on experience with objects that are usually left for equations. They will construct their own knowledge from meaningful experiences which leads to a constructivist way of learning.

Interests in immersive learning have grown in recent years as Virtual Reality and Augmented Reality (VAR) technology have become more accessible. Such digital technologies have revolutionized educational opportunities by allowing people to experience and interact with diverse environments. There are numerous ways in which VAR technology can be employed to deliver immersive learning experiences: augmented reality projections, virtual constructed 3D realities and interactive 360° videos. Also, Immersive Simulation-Based Learning (ISBL) have been successfully applied in numerous economy and engineering courses at Pennsylvania State University using SIMIO simulation software, (Nowparvar et al. 2022). Furthermore, the immersive simulations are developed with simulation software which does not incur any technology fee for academic and classroom use. Then students have the option to learn in low-immersion mode or in high-immersion mode via Virtual reality headset. The advantage of the low-immersion mode is that students can learn even remotely, from their home, without the need to buy their own expensive headset. Also the merge of game-based approaches and Virtual Reality environment is considered as a highly promising technique to enhance immersive learning (Checa and Bustillo 2020). Serious games are a well-known learner-centered approach to education widely used during the last decade. Merging serious games with immersive technology allows achieving high learning rates: user interactivity with the VAR environment. In fact, the use of games is the natural way to achieve high levels of interactivity.

Now, concretely, the most challenging points in incorporating ISBL in the ENIT Lab 4.0 are the requirement for teachers to undergo training and possess a significant understanding of the intricacies involved. Additionally, the expense of equipment can be substantial, and the shift to a more student-centered, project-based learning approach may not always be warmly received by faculty members as well as the engineering students.

4 Impact on Engineering Education

The aforementioned novel technologies and immersive learning can greatly improve knowledge retention compared with traditional educational methods; furthermore, it enhances critical thinking and empathy (Motyl and Filippi 2021). These competencies are of great interest to prepare the future ENIT students to deal with the green and digital transition. Thus, the objective of the present project is to enhance courses with original teaching content designed using this immersive approach to improve learning outcomes in environmental sustainability. Specifically, the project aims to prepare novel MOOC and ISBL modules using SIMIO simulation software dedicated to green and digital transition context such as, among other, sustainable supply chain, circular production and remanufacturing. With such immersive learning, the engineering students will have new opportunities to interact with assets, materials as well as realistic industrial environments in which their physical presence would be too expensive. This immersion will also encourage the collaborative construction of knowledge between different students which will offer more inclusive learning experience.

5 Conclusion

A novel trend in several prestigious engineering schools have gone beyond teaching their engineering students about additive manufacturing by integrating it into a hands-on, immersive learning method. This approach provides a more engaging and innovative experience. With its industry 4.0 laboratory, the National Engineering School of Tunis is on this path. However, the main challenge to be addressed is the need to have adequate equipment and a properly trained teaching staff.

Further work should be conducted on assessing the effectiveness of the AM as well as the upcoming ISBL learning approaches. Surveys should be conducted to evaluate these modules for teaching and learning engineering concepts within ENIT's teachers and students.

Acknowledgements. This project has received funding from the European Union's "Erasmus+ Capacity Building in the field of Higher Education" program under grant agreement No 2019–1949 / 001–001 (correspondent to the project shortly entitled "NePRev", "NExt Production REVolution").

Authors are grateful to SIMIO for providing the discrete-event simulation software under a grant from SIMIO LLC to the National Engineering School of Tunis.

References

Butt, J.: Exploring the interrelationship between additive manufacturing and Industry 4.0. Designs **4**(2), 13 (2020). https://doi.org/10.3390/designs4020013

Checa, D., Bustillo, A.: A review of immersive virtual reality serious games to enhance learning and training. Multimedia Tools Appl. **79**, 5501–5527 (2020). https://doi.org/10.1007/s11042-019-08348-9

Jaoua, A., Elarbi, Y., Jaoua, M.: Application of additive manufacturing for innovative design and development of reusable packaging. In: Proceedings of the 4th Euro-Mediterranean Conference for Environmental Integration (EMCEI-2022), Sousse, Tunisia, Recent Advances in Environmental Science from the Euro-Mediterranean and Surrounding Regions, Advances in Science, Technology & Innovation, Springer. (2022). In Press

Jaoua, A., Negri, E., Layeb, S.B., Ayed, Z.: Framework for smart online 3D bin packing using augmented reality. In: Proceedings of the 6th International Conference on Advanced Systems and Emergent Technologies (IC-ASET 2023), Hammamet, Tunisia, IEEE conference. (2023). In press

Marks, B., Thomas, J.: Adoption of virtual reality technology in higher education: an evaluation of five teaching semesters in a purpose-designed laboratory. Educ. Inf. Technol. **27**(1), 1287–1305 (2022). https://doi.org/10.1007/s10639-021-10653-6

Melo, G., Ercan, A., Kolter, M., Schleifenbaum, J.H.: Investigating the use of immersive technologies for additive manufacturing. In: Advances in Production Management Systems. Smart Manufacturing and Logistics Systems: Turning Ideas into Action: IFIP WG 5.7 International Conference, APMS 2022, Gyeongju, South Korea, September 25–29 2022, Proceedings, Part II. Springer Nature, Cham, pp. 469–476 (2022). https://doi.org/10.1007/978-3-031-16411-8_54

Motyl, B., Filippi, S.: Trends in engineering education for additive manufacturing in the industry 4.0 era: a systematic literature review. Int. J. Interact. Des. Manuf. (IJIDeM) **15**(1), 103–106 (2021). https://doi.org/10.1007/s12008-020-00733-1

Nowparvar, M., Ashour, O., Ozden, S., Knight, D., Delgoshaei, P., Negahban, A.: An assessment of simulation-based learning modules in an undergraduate engineering economy course. In: 2022 ASEE Annual Conference & Exposition (2022)

ISO ASTM 59900: Additive Manufacturing—General Principles Terminology. International Organization for Standardization, Geneva, Switzerland (2015)

Exploring Human-Cyber-Physical Systems in Additive Manufacturing: Insights into Human-Machine Collaboration

Anis Hamza(✉) and Noureddine Ben Yahia

Mechanical, Production and Energy Laboratory (LMPE), National School of Engineering of Tunis (ENSIT), University of Tunis, Avenue Taha Hussein Montfleury, 1008 Tunis, Tunisia
anis7amza@gmail.com

Abstract. The technology known as additive manufacturing, or 3D printing, has swiftly emerged as a revolutionary force that could transform the process of product design and manufacturing. This convergence of the physical and digital worlds has given rise to human-cyber-physical systems (HCPS) within additive manufacturing. HCPS involves seamless collaboration and interaction between humans, machines, and software, creating new possibilities for designing and producing products. Consequently, comprehending the intricacies of human-machine collaboration within HCPS becomes essential for unlocking the complete potential of additive manufacturing. Building upon our previous work, we aim to leverage our expertise in human-machine interaction to explore its application within the realm of additive manufacturing. This paper aims to investigate HCPS in Additive Manufacturing (AM) by providing valuable insights into the factors that contribute to effective collaboration between humans and machines. Through a comprehensive analysis of existing literature and case studies, we intend to shed light on the challenges and opportunities in this field, as well as identify key areas for future research and development.

Keywords: Additive manufacturing (AM) · Human-Cyber-Physical Systems (HCPS) · Human-Machine collaboration · Industry 5.0 · Internet of Things (IoT)

1 Introduction

As a perspective of our previous work (Anis Hamza et al. 2023 and 2021), our objective is to apply our expertise in the field of human-machine interaction to the domain of additive manufacturing. Industry 5.0 (Refer to Fig. 1) signifies an expansion of the existing industrial structure, aiming to maximize profits through product production and services, while shifting towards a human-centric, sustainable, and resilient industry. In essence, the manufacturing-oriented industry will strive to achieve a flexible and sustainable human-centered approach by automating production systems in response to increasingly complex global competition. Industry 5.0 evolves through the coexistence of individuals and data-driven technologies, facilitated by digital innovations like robots,

T. Mabrouki et al. (Eds.): AIAM 2023, LNME, pp. 92–99, 2024.
https://doi.org/10.1007/978-3-031-47784-3_12

Virtual Reality (VR), Augmented Reality (AR), and Artificial Intelligence (AI). Industry 5.0 can be categorized into six domains: individualized human-machine interaction, bio-inspired technologies and smart materials, digital twins and simulations, data transmission, storage, and analysis technologies, artificial intelligence (AI), and technologies for energy efficiency and renewable energy. Human-Centric Intelligent Manufacturing (HCIM) stands as a central value within Industry 5.0 (Yi Xiong et al. 2023 and Rahman MA et al. 2023), representing the latest phase of the industrial revolution. HCIM places significant emphasis on placing humans at the core of the manufacturing process and harnessing advanced technologies such as AI, robotics, and the Internet of Things (IoT) to enhance human capabilities and elevate the quality of work (Xun Xu et al. 2021).

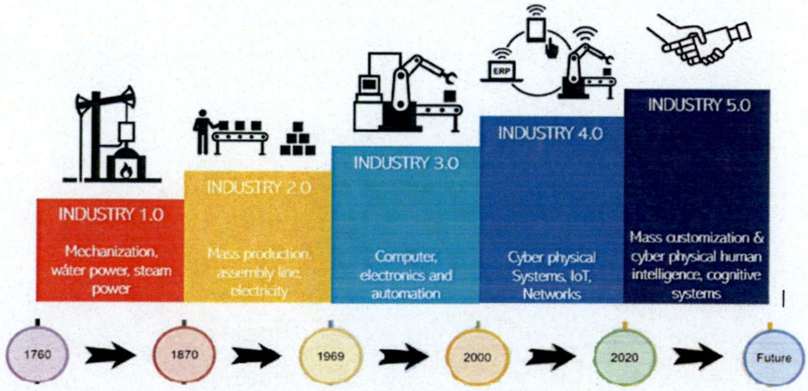

Fig. 1. Industrial revolutions in human history

In the realm of Human-Centric Intelligent Manufacturing (HCIM), human workers are no longer seen as passive operators who simply carry out predetermined tasks, but as active collaborators who work alongside machines and algorithms to achieve shared objectives. This necessitates a shift in the design and organization of manufacturing systems towards more flexible and adaptable production environments capable of seamlessly integrating human and machine intelligence. HCIM has garnered increasing attention in both academic and industrial spheres due to its potential benefits, including heightened productivity, improved quality, enhanced job satisfaction, and increased safety. By equipping workers with intelligent tools and technologies, HCIM can address challenges faced by traditional manufacturing, such as labor shortages, low efficiency, and repetitive tasks. Additionally, HCIM can cultivate innovation and creativity by enabling workers to focus on complex and value-added activities like problems solving, decision-making, and knowledge sharing (Mario Nardo et al. 2020).

Overall, HCIM presents a promising paradigm for the future of manufacturing, where the integration of human and machine intelligence opens up new possibilities for growth, sustainability, and social progress. The concept of Human-Centered Intelligent Manufacturing (HCIM) revolves around meeting the needs of humans within the manufacturing system and ensuring that intelligent manufacturing systems are developed with human factors in mind. This approach emphasizes the significance of considering the human element in the design, development, and implementation of intelligent manufacturing systems (Xiao Chen et al. 2021).

This study explores the concept of HCIM from various angles, including its background, meaning, human factors, technical systems, and practical applications. The findings indicate that HCIM represents a critical research direction for intelligent manufacturing and should be linked with relevant national strategies. It has should be regarded as a key concept for enterprise development and supported through enhanced research on human factors ergonomics in Intelligent Manufacturing Systems (IMS).

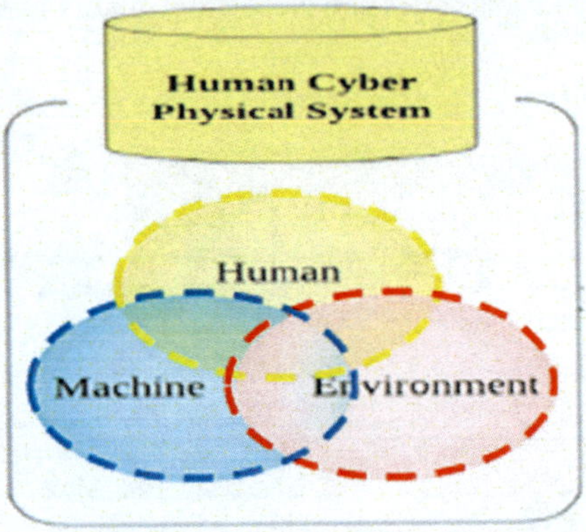

Fig. 2. The Hman–Cyber–Physical System (HCPS)

The study underscores the significance of human factors in the development of intelligent manufacturing systems, emphasizing that these factors should be taken into account throughout the entire system development lifecycle. It highlights that the success of intelligent manufacturing systems hinges on their ability to cater to human needs, requiring a design and implementation process rooted in a deep understanding of human behavior, cognition, and preferences (Guoquan Zhang et al. 2023). In conclusion, Human-Centric Intelligent Manufacturing (HCIM) emerges as a vital perspective and research direction for intelligent manufacturing, underscoring the importance of considering human factors in system development and advocating for a holistic approach to design that integrates human needs and preferences. The study provides recommendations for policymakers, enterprises, and researchers to advance the development and implementation of HCIM.

The theory of Human-Cyber-Physical Systems (HCPS) recognizes the interconnectedness and interdependence of humans, technology (cyber), and physical systems (machines, equipment, etc.) in the manufacturing domain (Refer to Fig. 2). Manufacturing's ultimate goal is to enhance human well-being, thus humans are an integral part of the manufacturing process (Lu. Yuqian et al. 2022) Building upon this theory, the concept and framework of Human-Cyber-Physical Interaction in Manufacturing (HCIM) have been established and investigated. HCIM aims to design and optimize manufacturing systems that facilitate effective interactions and collaborations between humans, technology, and physical systems to improve productivity, quality, and safety. HCIM considers various factors, including the physical layout of the production system, task allocation between human operators and machines, the use of automation and control systems, and the provision of appropriate training and support for human operators. In summary, HCIM offers a comprehensive approach to manufacturing design and optimization by acknowledging the interdependence of humans, technology, and physical systems. By fostering effective interactions and collaborations between humans and machines, HCIM has the potential to enhance the well-being of human operators, boost productivity, and optimize the overall manufacturing process.

2 Human-Machine Collaboration in Additive Manufacturing

Additive manufacturing, also known as 3D printing, offers several advantages that have revolutionized the manufacturing industry. These advantages include: Design freedom: One of the key benefits of additive manufacturing is the unprecedented design freedom it offers. Traditional manufacturing methods often impose limitations on design due to tooling constraints. However, with additive manufacturing, complex geometries and intricate designs can be easily produced, allowing for innovative and creative product designs. Customization: Additive manufacturing enables the production of highly customized products. This technology allows for individualized modifications and adaptations, catering to specific customer needs and preferences. Customization has become a crucial aspect of many industries, such as medical devices, where personalized solutions can significantly improve patient outcomes.

Waste reduction: Additive manufacturing is inherently more sustainable compared to traditional manufacturing methods. It operates on an additive principle, building objects layer by layer using only the necessary materials. This approach minimizes material waste, as it avoids the need for subtractive processes that generate excess scrap material. Additionally, the ability to create complex designs with optimized structures can reduce material consumption further. Rapid prototyping: Additive manufacturing excels in rapid prototyping, allowing for quick iterations and modifications during the product development stage. Traditional manufacturing methods often involve lengthy and costly tooling processes, making design changes time-consuming and expensive. With additive manufacturing, prototypes can be rapidly produced, tested, and refined, accelerating the product development cycle. Decentralized manufacturing: Additive manufacturing has the potential to decentralize manufacturing processes. It reduces the reliance on centralized production facilities, as 3D printers can be located closer to the point of use. This decentralization brings numerous advantages, including reduced transportation costs, faster production turnaround times, and the ability to produce on-demand, eliminating the need for large inventories.

Overall, additive manufacturing (AM) has transformed the manufacturing landscape by offering design freedom, customization, waste reduction, rapid prototyping, and the potential for decentralized manufacturing. These advantages have opened up new possibilities across various industries and are continuing to drive innovation and advancement in manufacturing processes.

Human-machine collaboration (Wang Baicun et al. 2022) constitutes a vital element within additive manufacturing (AM) systems (Refer to Fig. 3). AM technologies, such as 3D printing, rely on computer-controlled machines working in conjunction with human operators to fabricate intricate 3D objects. These machines rely on input from human operators to ensure the success of the process and to make real-time adjustments for process optimization. Human-machine collaboration in AM offers several notable advantages. Firstly, it enables greater flexibility and customization. Human operators can swiftly adapt the process based on their expertise and knowledge, leading to faster and more efficient production. Furthermore, human operators can identify potential issues or defects and promptly address them, minimizing waste and enhancing the overall quality of the final product. Additionally, human-machine collaboration enhances safety as human operators can oversee the process, ensure adherence to safety protocols, and proactively identify and mitigate potential hazards. Nonetheless, there are challenges associated with human-machine collaboration in AM. One primary challenge revolves around the need for training and expertise (Faccio M. et al. 2023). Human operators must possess in-depth knowledge of the AM process and effectively work alongside the machines, necessitating extensive training and continuous education to equip operators with the requisite skills and knowledge. Another challenge lies in the potential for errors and mistakes. While human operators can swiftly adjust the process, there is a risk of introducing errors if they are not meticulous, resulting in wasted materials, time, and resources.

To summarize, human-machine collaboration emerges as a pivotal component within AM systems, enabling greater flexibility, customization, and safety. However, successful collaboration necessitates extensive training and expertise. As AM technology continues to advance, the importance of human-machine collaboration in optimizing the process and enhancing final product quality is expected to grow.

There are several typical application scenarios where Human-Machine Collaboration is crucial in Additive Manufacturing (AM) productions, including: First, Prototyping: Industries such as automotive, aerospace, and medical device manufacturing often rely on AM for prototyping. In this scenario, human operators closely collaborate with machines to ensure that prototypes meet the required specifications. Second, Customization: AM facilitates the production of customized products tailored to specific customer needs. Human operators play an essential role by engaging with customers to determine requirements and adjusting the AM process accordingly. Third, Small batch production: AM efficiently produces small batches of parts or components. Human-machine collaboration is necessary in this scenario to optimize the process for the batch size, ensuring quality and consistency through necessary adjustments. Fourth, Repair and maintenance: AM can generate replacement parts for out-of-production machinery or equipment. Human operators are critical in this scenario as they closely collaborate with machines to ensure compatibility and adherence to specifications. Finally, Education and research: AM serves as a valuable educational and research tool, enabling students and researchers to explore new applications and technologies. Human-machine collaboration is vital as operators work alongside students and researchers to develop processes, techniques, and optimize AM for specific applications.

In conclusion, human-machine collaboration plays a vital role across various application scenarios in additive manufacturing. Whether it involves prototyping, customization, small batch production, repair and maintenance, or education and research, human

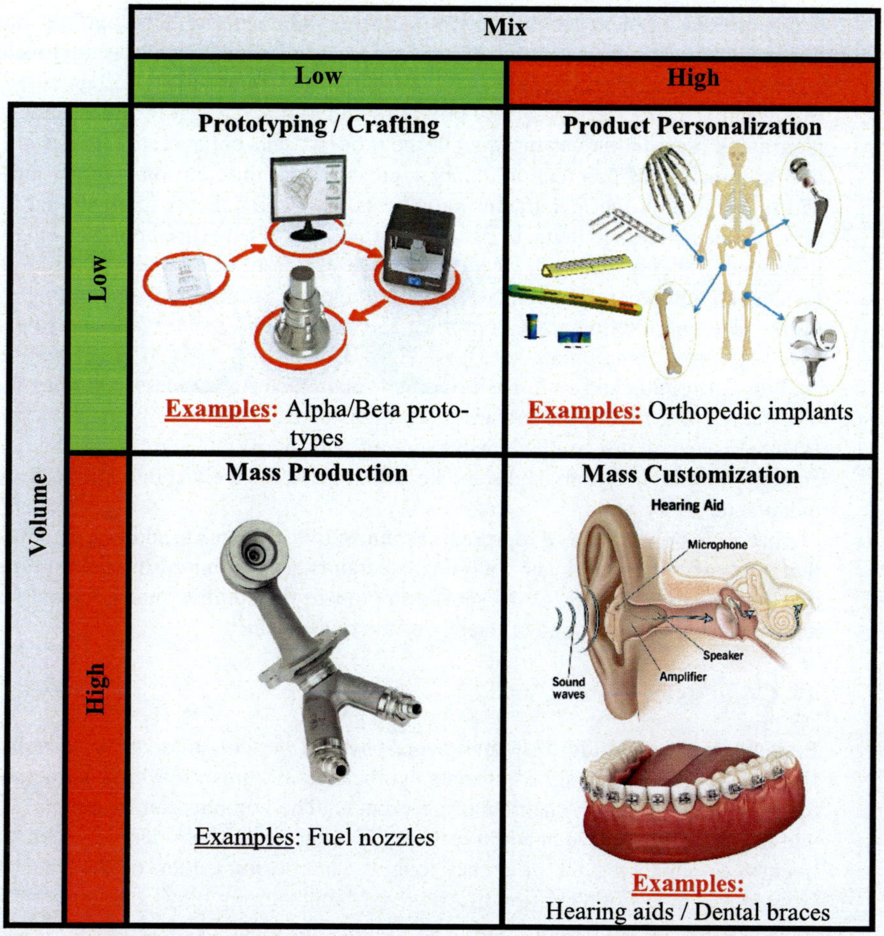

Fig. 3. Typical application scenarios for Additive Manufacturing (AM) productions that need Human-Machine Collaborations (HMC).

operators ensure that the AM process is optimized for specific applications and produces final products meeting the required specifications.

3 Concept of Human-Machine Collaborative Additive Manufacturing (HMCAM)

Human-Machine Collaborative Additive Manufacturing (HMCAM) encompasses the interaction between human operators and additive manufacturing systems, aiming to optimize the overall additive manufacturing process. HMCAM acknowledges the significant role of human operators in manufacturing and utilizes their expertise and knowledge to enhance the efficiency and quality of additive manufacturing. It recognizes that additive manufacturing is not entirely autonomous and requires human oversight and optimization. Human operators play an active role in Human-Machine Collaborative Additive Manufacturing (HMCAM) by providing input, making real-time adjustments, and identifying potential issues throughout the process. This collaborative approach, based on the principle of co-creation, enables greater flexibility, customization, and adaptability throughout the manufacturing process. Key components of HMCAM include: First, Human expertise: Human operators bring critical expertise and knowledge to the manufacturing process, such as identifying design issues and optimizing the process for desired outcomes. Second, Machine intelligence: Additive manufacturing systems utilize machine intelligence to optimize the process and make real-time adjustments based on input from human operators. Third, Data analytics: HMCAM leverages data analytics to monitor and optimize the additive manufacturing process, including the identification of potential issues and necessary adjustments. Finally, Training and education: HMCAM necessitates comprehensive training and education to ensure that human operators possess the necessary skills and knowledge to effectively collaborate with additive manufacturing systems.

In conclusion, HMCAM represents an innovative approach to additive manufacturing that recognizes the significance of human operators. By combining human expertise with machine intelligence, HMCAM seeks to optimize the additive manufacturing process and enhance the quality and efficiency of the final product.

4 Conclusion

In conclusion, the study of human-cyber-physical systems in additive manufacturing has yielded valuable insights into the significance of human-machine collaboration in optimizing the additive manufacturing process. The combination of human expertise and machine intelligence in additive manufacturing systems has demonstrated its effectiveness in enhancing the efficiency, quality, and customization of the final product. Human-Machine Collaboration in Additive Manufacturing plays a pivotal role in various scenarios, including prototyping, customization, small batch production, repair and maintenance, and education and research. By working alongside machines, human operators can provide input, make real-time adjustments, and identify potential issues during the process.

The concept of Human-Machine Collaborative in Additive Manufacturing (HMCAM) acknowledges the importance of human operators in the manufacturing process and aims to harness their expertise and knowledge to optimize additive manufacturing. HMCAM is grounded in the principle of co-creation, where human operators collaborate with machines to achieve the desired outcomes.

Moving forward, further research and development in the field of HMCAM will be essential to fully unlock the potential of human-machine collaboration in additive manufacturing. This will necessitate comprehensive training and education to equip human operators with the necessary skills and knowledge to effectively work with additive manufacturing systems. Additionally, continued investment in machine intelligence and data analytics will be critical to optimize the additive manufacturing process and enhance the quality and efficiency of the final product.

References

Hamza, A., Ben Yahia, N.: Artificial neural networks controller of active sus-pension for ambulance based on ISO standards. Proceed. Instit. Mech. Eng. Part D: J. Automobile Eng. **237**(1), 34–47 (2023). https://doi.org/10.1177/09544070221075456

Yi, X., Yunlong, T., Samyeon, K., Rosen, D.W.: Human-machine collaborative additive manufacturing. J. Manuf. Syst. **66**, 82–91 (2023)

Zhang, G., Wang, Y., Chen, Z., Xuguang, X., Dong, K., Xiong, Y.: Robot-assisted conformal additive manufacturing for continuous fibre-reinforced grid-stiffened shell structures. Virt. Phys. Prototyping **18**, 1 (2023)

Rahman, M.A., et al.: Review of intelligence for additive and subtractive manufacturing: current status and future prospects. Micromachines **14**(3), 508 (2023)

Faccio, M., Granata, I., Menini, A., et al.: Human factors in cobot era: a review of modern production systems features. J. Intell. Manuf. **34**, 85–106 (2023)

Yuqian, L., et al.: Outlook on human-centric manufacturing towards Industry 5.0. J. Manuf. Syst. **62**, 612–627 (2022)

Hamza, A., Ben Yahia, N.: Heavy trucks with intelligent control of active suspension based on artificial neural networks. Proceed. Institut. Mech. Eng. Part I: J. Syst. Control Eng. **235**(6), 952–969 (2021). https://doi.org/10.1177/0959651820958516

Xu, X., Yuqian, L., Birgit, V.-H., Lihui, W.: Industry 4.0 and Industry 5.0 inception, conception and perception. J. Manuf. Syst. **61**, 530–535 (2021)

Xiao, C., Eder, M.A., Shihavuddin, A., Zheng, D.: A human-cyber-physical system toward intelligent wind turbine operation and maintenance. Sustainability **13**(2), 561 (2021)

Mario, N., Forino, D., Murino, T.: The evolution of man–machine interaction: the role of human in Industry 4.0 paradigm. Prod. Manufact. Res. **8**(1), 20–34 (2020)

Baicun, W., Pai, Z., Yue, Y., Albert, S., Lihui, W.: Toward human-centric smart manufacturing: a human-cyber-physical systems (HCPS) perspective. J. Manufact. Syst. **63**, 471–490 (2022)

Simplification of an Additive Manufacturing Machine Implementation Using Its CAD Model and Mixed-Reality

Mina Ghobrial[1](✉), Philippe Seitier[1], Michel Galaup[2], Pierre Lagarrigue[3], and Patrick Gilles[1]

[1] ICA, Institut National des Sciences Appliquées de Toulouse, 135 Av. de Rangueil, 31400 Toulouse, France
{ghobrial,philippe.seitier,patrick.gilles}@insa-toulouse.fr

[2] EFTS, Institut National Universitaire Champollion, Pl. de Verdun, 81000 Albi, France
michel.galaup@univ-jfc.fr

[3] ICA, Institut National Universitaire Champollion, Pl. de Verdun, 81000 Albi, France
pierre.lagarrigue@univ-jfc.fr

Abstract. This study presents the development and evaluation of a Mixed Reality (MR) tool called HoloPrint, designed to assist students in learning how to use a Filament Deposition Machine (FDM) for 3D printing. The HoloPrint tool combines a MR headset with a pedagogical scenario based on teacher feedback and is tailored to the expected competency profiles of students. The proof of concept phase focuses on understanding how users adopt this novel way of interacting with a combined virtual and real environment while accomplishing the scenario's objectives. A homogenous population of 81 first-year engineering students at INSA Toulouse was selected to participate in the study. The experiment was conducted during practical work sessions with 8 classes, each consisting of 10 students. The study evaluated the success rate of students using HoloPrint and assessed the tool's usability and usefulness through the System Usability Scale (SUS) questionnaire. Results indicate a high success rate of 98.8% for students using HoloPrint to print objects without assistance. The average SUS score for the tool was 79.69, indicating "Good" acceptability and nearing "Excellent." Furthermore, prior experience with 3D printers or VR headsets did not significantly impact the success rate or SUS scores, suggesting the tool's effectiveness in aiding both novice and experienced students. In conclusion, the study demonstrates the potential of Mixed Reality as an effective and user-friendly tool for training students on a simple additive manufacturing machines, it enables the researchers to conduct further experimentation on more complex additive manufacturing machines.

Keywords: Additive manufacturing · 3D printer · augmented reality · training · acceptability

T. Mabrouki et al. (Eds.): AIAM 2023, LNME, pp. 100–106, 2024.
https://doi.org/10.1007/978-3-031-47784-3_13

1 Introduction

Metal additive manufacturing has revolutionized the manufacturing industry, allowing for the production of complex and intricate metal parts that were previously impossible to manufacture. However, with the increased complexity of these machines comes higher operating and maintenance costs, as well as greater safety concerns especially with the introduction of highly toxic metallic powder.

In order to address these challenges, the industry may turn to immersive technologies such as Augmented Reality (AR) and Mixed Reality (MR) to help with operator training, reduce costs, and mitigate safety risks. AR and MR technologies offer a unique and interactive way for operators to learn how to use and maintain these complex machines, while also reducing the risk of accidents and errors.

By simulating the operation of the machines in a virtual environment, operators can practice and learn without the need for expensive, time-consuming, and potentially dangerous on-the-job training. Additionally, image processing techniques can help operators identify and diagnose potential maintenance issues before they become more serious and costly problems.

This article will explore this idea by developing and testing an immersive tool on a group of first-year engineering students a MR scenario using a Hololens 2 headset to train first year engineering students on using a 3D printer that is a relatively simple and safe additive manufacturing machine.

2 Literature Review

J. Butt (Butt 2020) discusses the use of Augmented Reality (AR) as a tool for advanced manufacturing systems. The author mentions that advanced manufacturing machines vary in size, feedstock usage, and product properties, and that controlling these machines is a challenging aspect that can benefit from the type of digital transformation associated with Industry 4.0. The author provides examples of how AR can be used for visualization and control of advanced manufacturing machines, such as the development of bi-directional AR interfaces for 3D printers, an interactive fabrication system that provides an in-situ modeling experience that can offer tangible feedback to the user, and a scan-based method for real-time monitoring of a 3D printer. Additionally, the author notes that AR can be used to superimpose virtual models over physical ones to assess geometrical errors and deviations. Gonzalez-Franco *et al.* (Gonzalez-Franco et al. 2017) explore the use of Mixed Reality (MR) for complex manufacturing training and present the results of a user study measuring post-training knowledge retention and interpretation scores. The study found that MR training showed equivalent knowledge retention to conventional face-to-face training, but no significant differences or equivalences for knowledge interpretation. The results of the study by Ostrander *et al.* (Ostrander et al. 2020) suggest that utilizing both interactive and passive Virtual Reality (VR) may be used to effectively teach introductory concepts of Additive Manufacturing (AM); they also found advantages to using interactive VR for improving AM self-efficacy. On the other hand Mogessie *et al.* (Mogessie et al. 2020) have actually developed an intelligent training tutor using VR to train users on a specific AM machine - the EOS M290. The implementation of MR training in the industry has several benefits, including reduced costs and elimination of security issues and operational hazards. However, professional trainers would still be needed, therefore, only part of the overhead training costs would be reduced. The study provides evidence of how the integration of existing metaphors for collaborative work can be implemented in immersive MR.

This paper explores the utility and usability of Mixed Reality in additive manufacturing and represents a continuation of work by Seitier *et al.* on the use of Augmented Reality (Seitier & Gilles 2021) and Mixed Reality (Seitier et al. 2022) in education and training.

3 Experimental Setup

The aim of this research is to assess the effectiveness of Mixed Reality technology in the implementation of an additive manufacturing machine, and to measure the benefits of this approach during operation. To evaluate the advantages gained through the use of Mixed Reality, a System Usability Scale (SUS) questionnaire (Brooke 1995) is used.

The specific additive manufacturing machine selected for this implementation is an Ultimaker 3 3D printer, and the Mixed Reality device used is a Hololens 2 headset, as illustrated in Fig. 1-a.

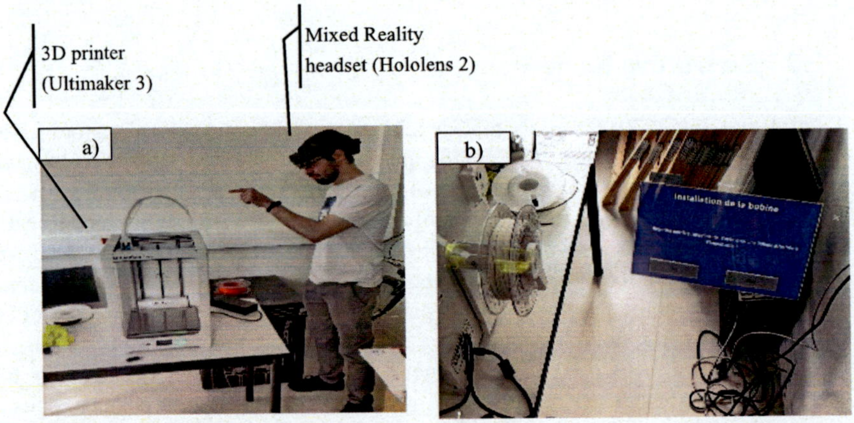

Fig. 1. a) Experimental setup hardware, b) A filament coil installation scenario step.

This study aims to develop a proof of concept for a tool called HoloPrint, which can assist students in getting started using a Filament Deposition Machine (FDM). The machine is relatively simple and safe to use, making it an ideal starting point to test out the HoloPrint tool. HoloPrint is a combination of a Mixed Reality headset and a pedagogical scenario based on feedback from teachers and the Ultimaker's CAD model. The tool is designed to respond to the expected competency profiles of students.

During the proof of concept phase, the study focuses on the modalities of use of the device to address the fundamental issues of using Mixed Reality to tackle the addressed question. The goal is to understand how users adopt this new way of acting and perceiving an environment combining virtual and real objects and how they can complete the scenarii's objectives.

By testing the HoloPrint tool in this study, the researchers hope to understand how students can better interact with 3D printers using Mixed Reality. This will enable the research team to gain insights into how to improve the tool to make it more effective for users. The study aims to lay the foundation for the development of similar tools that can be used in different professional situations, ultimately contributing to the improvement of learning experiences for students.

The group of individuals being studied consists of 81 students who are in their first year at INSA, a French engineering school in Toulouse. These students were selected based on specific and well-defined criteria, making them a relatively homogenous population. Furthermore, they have all followed a 30h mechanical design course before. The experiment was conducted during a practical work class in the same workspace and work setup. There were 8 classes in total, with each class consisting of 10 students. Table 1 provides detailed information about the main composition of the population of students who participated in this experiment.

Table 1. Population composition

Criterion	Ratio (%)
Male	53
Female	46
Non-binary	1
New to XR usage	52
New to 3D printer usage	89

In order to create the scenario for this experiment, several assumptions were made about the students being studied. It was assumed that they were all under the age of 20, had experience with new technologies, and owned a smartphone. This presumption was later verified: a significant majority of the students, over half, had past experience with an immersive, head-mounted display, most often for gaming or controlling drones via a VR headset. Interestingly, none of the students had any experience with a Mixed Reality head-mounted display before. The potential benefits of incorporating gaming elements into the scenario were also considered, as they could heighten the students' involvement in the task. The intention was to ensure that the usage of a new technology, a MR headset, wouldn't distract the students, therefore, a 3D drawing application was used as an acculturation step. This allowed the students to become familiar with the Hololens 2 before using it with the scenario. The scenario itself is linear and consists of tasks that must be completed in order. For example, the students are asked whether or not they want to change the spool. If the task has already been completed, the student is brought back to the main menu. Since the scenario requires the use of the head-mounted display, it was considered to be a single tool along with the scenario itself.

The success of the test is measured by the student's ability to launch a print process without any assistance. A standardized System Usability Scale (SUS) questionnaire is used to evaluate the usefulness and usability of the developed tool, and it consists of 10 questions with five response options going from "Strongly agree" to "Strongly disagree". The questions are the following:

1. I think that I would like to use this application frequently
2. I found the system unnecessarily complex
3. I thought the system was easy to use
4. I think that I would need the support of a technical person to be able to use this system
5. I found the various functions in this system were well integrated
6. I thought there was too much inconsistency in this system
7. I would imagine that most people would learn to use this system very quickly
8. I found the system very cumbersome to use
9. I felt very confident using the system
10. I needed to learn a lot of things before I could get going with this system

4 Results and Interpretation

The SUS methodology is used to compute the scores of all participants on a scale ranging from 0 to 100, as represented in Bangor *et al.*'s study (Bangor et al. 2009). The cumulative evaluation is performed by considering the mean scores of all involved participants.

In this study, it was found that the success rate of students using the HoloPrint tool was high, with 98.8% of the students successfully printing an object without any external help (80 students out of 81). Fascinatingly, the success rate didn't show a noticeable difference between students who had experience with a 3D printer and those who didn't. This suggests that the tool was equally effective in fostering autonomy in both beginner and relatively experimented students. The average score for the tool stood at 79.69 (as seen in Fig. 2), which, as per Bangor *et al.*'s research (Bangor et al. 2009), signifies

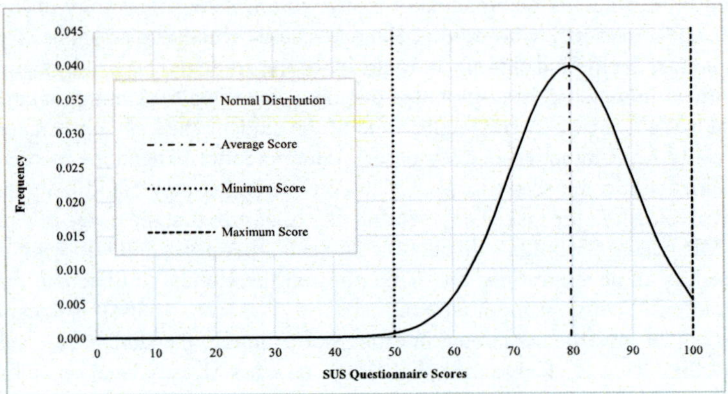

Fig. 2. SUS Scores normal distribution

the tool's acceptability was considered "Good," nearing "Excellent." The calculated standard deviation was 10.10%, which implies that most students had a nearly uniform level of acceptability for the Proof of Concept (PoC). The considerable interest among students, noted after their initial use of the tool, points to the fact that their fascination wasn't merely triggered by the tool's newness.

The tool was also found to have a significant helping potential, with 95% of students indicating that they did not need to learn a lot before they could start using the application. This finding highlights the tool's user-friendliness and its ability to assist students without requiring extensive prior knowledge or skills. Furthermore, the use of an immersive head-mounted display prior to the experiment did not significantly affect the success rate or overall score obtained, suggesting that the HoloPrint tool was not dependent on prior immersive technology experience.

Overall, the results of this study suggest that the HoloPrint tool is an effective and user-friendly tool for teaching 3D printing to students, regardless of their prior experience with 3D printers or immersive technology. The tool's high success rate, acceptability, and helping potential make it a valuable addition to the classroom for both novice and experienced students.

5 Conclusion

This research set out to explore the application of Mixed Reality alongside additive manufacturing machinery, and assess the effectiveness of a specially developed Mixed Reality tool. Comprised of a Hololens 2 Augmented Reality headset and a designed scenario, this tool was put to the test with a homogeneous group of 81 first-year INSA Toulouse students, who were split into eight practical section groups.

The students were introduced to the tool in a hands-on workshop, initially engaging in a 3D drawing game to get a grasp of the technology. Subsequently, they used the tool for guidance while operating a 3D printer, and concluded with filling out an individual SUS questionnaire to share their impressions of the experience.

The results of the study indicate that the Mixed Reality tool was well received by the students, who found it easy to learn and use. With the exception of one student, all of the participants were able to successfully launch a 3D print, indicating that the tool was effective. Furthermore, the results of the SUS questionnaire showed that the students' responses were homogeneous and normally distributed around the mean value of 76.69% with a low standard deviation of about 10%, indicating that the tool was acceptable according to Bangor *et al.* (Bangor et al. 2009).

The study also identified several areas for improvement, including the possibility of using a smartphone as a medium to visualize the developed scenario, and the potential for transmitting skills using this personal technology approach.

To conclude, this study demonstrates the potential of Mixed Reality as a tool for training on a relatively simple additive manufacturing machine, and provides evidence of the efficacy of the Mixed Reality tool that was developed. The study also identifies areas for future research, and suggests that Mixed Reality has a role to play in facilitating the adoption of complex additive manufacturing technologies.

References

Bangor, A., Kortum, P., Miller, J.: Determining what individual SUS scores mean : adding an adjective rating scale. J. Usability Stud. **4**(3), 114–123 (2009)

Brooke, J.: SUS: a quick and dirty usability scale. In: Usability Evaluation in Industry 189 (1995)

Butt, J. (2020). Exploring the interrelationship between additive manufacturing and industry 4.0. Designs **4**(2), 13 (2020). https://doi.org/10.3390/designs4020013

Gonzalez-Franco, M., et al.: Immersive mixed reality for manufacturing training. Front. Robot. AI **4**, 3 (2017). https://doi.org/10.3389/frobt.2017.00003

Mogessie, M., Wolf, S.D., Barbosa, M., Jones, N., McLaren, B.M.: Work-in-progress—A generalizable virtual reality training and intelligent tutor for additive manufacturing. In: 2020 6th International Conference of the Immersive Learning Research Network (iLRN), pp. 355-358 (2020). https://doi.org/10.23919/iLRN47897.2020.9155119

Ostrander, J.K., Tucker, C.S., Simpson, T.W., Meisel, N.A.: Evaluating the use of virtual reality to teach introductory concepts of additive manufacturing. J. Mech. Des. **142**(5), 051702 (2020)

Seitier, P., Gilles, P.: Réalité augmentée en enseignement : L'outil ou la science ? 17ème colloque national S-mart AIP-PRIMECA (2021, mars). https://hal.science/hal-03296135

Seitier, P., Gilles, P., Boudier, V., Galaup, M., Lagarrigue, P.: Getting started procedure of a NC machine simplified by the use of a mixed-reality training scenario. MATEC Web Conf. **368**, 01018 (2022). https://doi.org/10.1051/matecconf/202236801018

A CAD-Based Method for the Measurement of AM Parts Accuracy Considering the Build Orientation

Ahmed Elayeb[1](✉), Anis Korbi[2], Riadh Bahloul[2], Farhat Zemzemi[1], Mehdi Tlija[2], and Borhen Louhichi[1]

[1] Mechanical Laboratory of Sousse, (LMS, ENISo), National Engineering School of Sousse, University of Sousse, 4023 Sousse, Tunisia
elayebahmed@gmail.com, borhen.louhichi@etsmtl.ca

[2] National Engineering School of Monastir, Mechanical Engineering Laboratory (LGM, ENIM), University of Monastir, 5019 Monastir, Tunisia

Abstract. In Additive Manufacturing (AM), several printing processes such as Fused Deposition Modelling (FDM) can deliver complex parts and shapes in short building time using different types of materials. However, the quality and the accuracy of the printed parts might vary according to several print input parameters. In this regard, the present paper aims to study and quantify the effect of the build orientation parameter on the dimensional accuracy of printed parts in FDM 3D printing. An example of a part prototype is manufactured based on the FDM technique and considering different build orientations. Points cloud of the printed prototype is extracted using the 3D scan method. The realistic CAD models of parts features are determined considering the point clouds resulting from the scan process. The dimensional accuracies of the reconstructed parts features are computed according to different build orientations. The developed method can be considered a preliminary decision support tool that assists designers and 3D printing manufacturers in the investigation of the build orientation impact on the accuracy of the printed parts features.

Keywords: AM · FDM · 3D printing · Build orientation · Dimensional accuracy · Points cloud · Reconstructed parts

1 Introduction

The first applications of 3D printing were in the area of rapid prototyping, which considered as a fundamental process to produce prototype models. Among all the AM technologies, FDM printing method is widely used for manufacturing or engineering purposes (Wong and Hernandez 2012) and considered as an economical process (Haidiezul et al. 2018) when contrasted with other techniques. The FDM printing of mechanical parts depends on many conditions. Creating an accurate and dimensionally stable part have always been a big challenging of FDM printing technology and depends on the choice of manufacturing parameters (Zharylkassyn et al. 2021). In the past, several research

T. Mabrouki et al. (Eds.): AIAM 2023, LNME, pp. 107–113, 2024.
https://doi.org/10.1007/978-3-031-47784-3_14

works were developed for the investigation of the impact of the manufacturing orientation in AM on the support material (Leary et al. 2013). However, the establishment of approaches for evaluation of the build orientation impact on dimensional and geometrical accuracies of the part features remains a complex task and is still among the major actual challenges of researchers in the AM field. In this regard, this work presents a new method for the prediction of the effect of the part orientation in AM on dimensional accuracy of printed part features.

2 State of the Art

In AM, the issue of dimensional stability becomes significant when printed components are proposed for assembly and customary use (Haghighi et al. 2017). Several research works are proposed in the state of the art for the study of the dimensional accuracy of parts in 3D printing considering different manufacturing parameters such as the built orientation. Wang et al. (2007) demonstrated that the dimensional accuracy of printed parts is affected by the build orientation and depositing thickness. The deposition patterns and the choice criteria of the optimal part orientation are the main causes of the differences in dimensional accuracy of final manufactured parts according to different printing directions. The Taguchi method was used by Sood et al. (2009) to detect the impact of FDM printing parameters such as the build orientation on the resulted dimensional accuracy of printed ABSP400 component. The above study proved that the dimensional errors of resulted parts are significant according to the Z-axis of the printing machine. Górski et al. (2017) determined the relation between with dimensional precision and repeatability of products in FDM printing technology considering the part orientation as the primary parameter. Singh et al. (2013) proved that the part position on the 3D printing machine has a significant effect on the dimensional accuracy and mechanical properties of manufactured components. The authors found that there is a small increment in dimensional accuracy error when progress along Y axis in the build. Elayeb et al. (2022) investigated the impact of elementary cylindrical deposit material model and the part orientation on dimensional feature errors by comparing a rectangular pocket dimensions in a part built along 3 orientations (X, Y and Z axis). In another work, Elayeb et al. (2023) also studied the dimensional accuracy through oblong elementary deposit material model and along different build orientations. The impact of these parameters are investigated through the case study of a USB 2.0 plug in order to decide the best build orientation affording the least dimensional errors.

3 Determination of the Build Orientation Effect on the Dimensional Accuracy

Several steps are followed and applied in order to determine the effect of the build orientation in FDM printing technology on the resulting dimensional accuracy of parts. A simple part model P1 illustrates the different steps of the approach. The main steps are as following:

3.1 3D Printing of the Part P1

The part P1 is first modelled in CAD using the SolidWorks Software. The dimensions and shape of P1 are presented in the Fig. 1. Rn is the nominal radius of each.

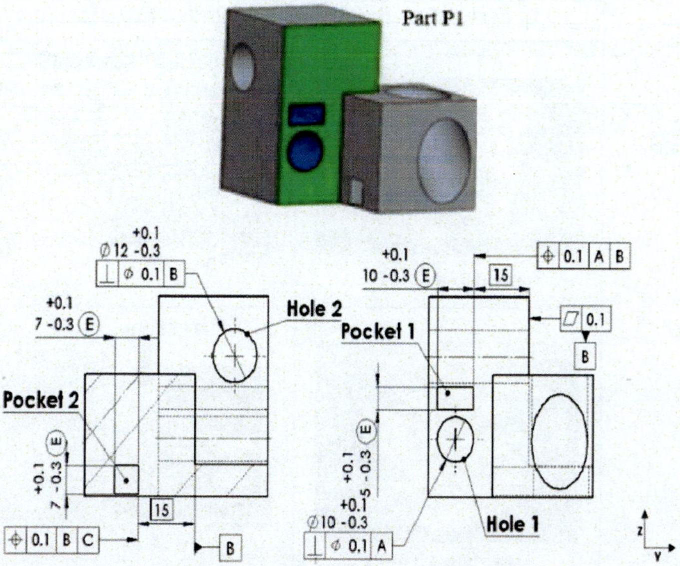

Fig. 1. Drawing of the part P1

Cylindrical hole. Lp and Hp are the nominal width and the Height of each rectangular pocket respectively.

After the CAD modelling of P1, the part is sliced and printed using the 'Raise3D Pro2' printer. The manufacturing parameters are presented in Table 1.

Table 1. The values of the 3D printing parameters.

Parameters	Value
Printing speed	60 mm/s
Density	100%
Melting T°	205°
Bed T°	60°
Nozzle Diameter	0.4 mm

The orientations of P1 during the 3D printing and the corresponding deposited material directions are illustrated in Table 2.

Table 2. Orientations of P1 and deposited material

The part orientations	Deposited material direction
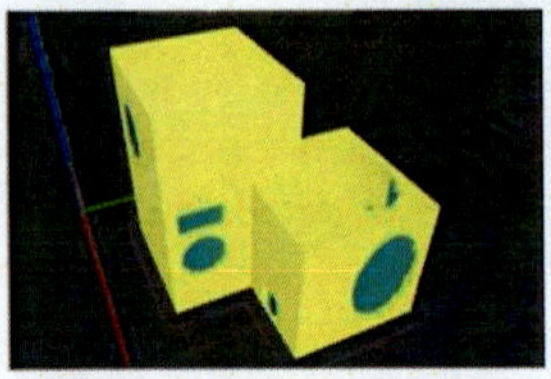 Orientation 1 of P1	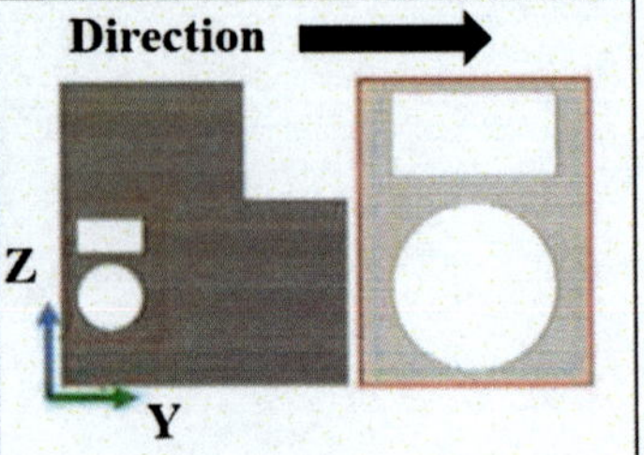
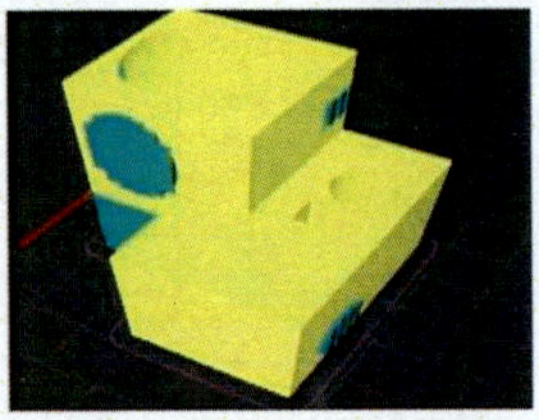 Orientation 2 of P1	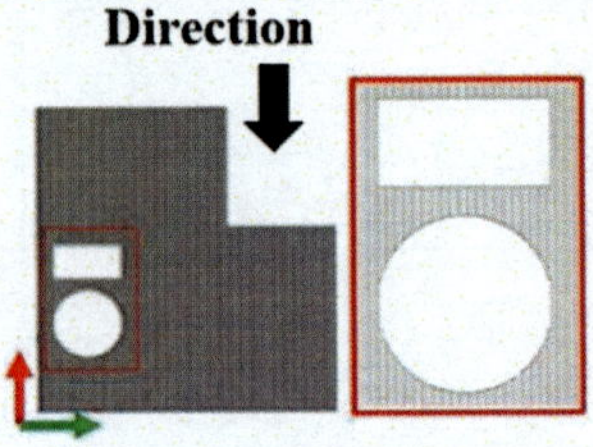

3.2 3D Scan and Reconstruction of the Part P1 Features

After getting the program results, the three airfoils are designed as showed in Fig. 2 The printed part P1 is first scanned using the 'EinScan-Pro+' scanner. After that, the resulted point's clouds of the scanned part are used to process for the reconstruction of the realistic part features based on the Levenberg Marquardt Algorithm (LMA) is applied. In the case of cylindrical holes, the outputs of the LMA are the center, the axis and the radius of the average cylinder that passes through the maximum number of scanned points (Fig. 2).

The realistic geometries of rectangular pockets are obtained thorough the reconstruction of all surfaces of the above features considering the scanned points of P1.

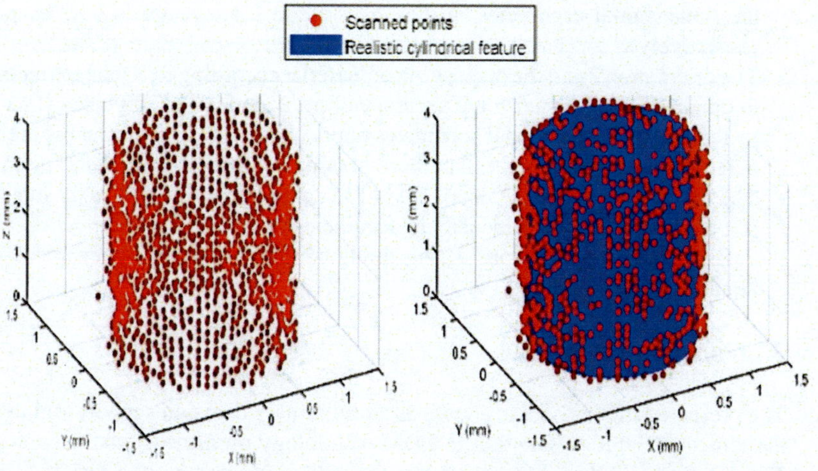

Fig. 2. Graphical illustration of a cylindrical hole reconstruction

4 Dimensional Accuracies of Cylindrical and Rectangular Features According to Different Build Orientations

After the reconstruction process, the dimensional accuracies of cylindrical holes and rectangular pockets of the part P1 are determined as presented in Table 3

Table 3. Dimensional accuracies of cylindrical features

Dimensional accuracies of the cylindrical hole 1 (All dimensions are in mm)				
Build orientation	Nominal radius Rn	Realistic radius Rh	Dimensional error $e = \lvert R_n - R_h \rvert$	Dimensional accuracy $E_{cyl}(\%) = 100(1 - \frac{e}{R_n})$
Orient 1	5	5,035	0,035	99.3%
Orient 2	5	5,062	0,062	98.76%
Orient 1	6	6,22	0,14	97,66%
Orient 2	6	6,14	0,22	96,33%

Results Analysis

The results of Table 3 show that:

- The orientation 1 and the disposition of material according to Y axis are recommended to obtain the best dimensional accuracy of printed holes (As presented in Table 3, the dimensional accuracies of the holes 1 and 2 are equal to 99.3% and 97.66% respectively).

- The orientation 2 and the disposition of material according to X axis are recommended to obtain the best dimensional accuracy of printed rectangular pockets (As presented in Table 3, the dimensional accuracies of pockets 1 and 2 according the the width and height locations are bigger than those obtained considering the build orientation 1
- The orientation 2 and the deposition of material according to X axis should be chosen to obtain the best dimensional accuracies of both holes and pockets. The above decision is performed considering the numerical values of dimensional accuracies already shown in Table 3.

5 Conclusion

The presented method is considered as a preliminary decision support tool to determine the effects of build orientation in FDM technology on dimensional accuracy of parts. The 3D scan and the LMA algorithm are used to reconstruct the realistic parts and to measure dimensional accuracies of features according to chosen parts orientations. The results present an improvement of the dimensional accuracy according to the x axis orientation. Further improvements should be performed in the future to consider both dimensional and geometrical errors of parts feature in the proposed method.

References

Elayeb, A., Korbi, A., Tlija, M., Louhichi, B.: Optimal part orientation in the additive manufacturing process based on tolerancing. In: Bouraoui, T., et al. (eds.) CoTuMe. LNME, pp. 217–225. Springer, Cham (2022). https://doi.org/10.1007/978-3-030-86446-0_29

Elayeb, A., Eltaief, A., Korbi, A., Tlija, M., Louhichi, B.: Printing orientation selection based on the dimensional errors modelling in additive manufacturing. In: Walha, L., et al. (eds.) Design and Modeling of Mechanical Systems – V, pp. 195–205. Springer, Cham (2023). https://doi.org/10.1007/978-3-031-14615-2_23

Górski, F., Wichniarek, R., Kuczko, W., Zawadzki, P., Buń, P.: Dimensional accuracy of parts manufactured by 3D printing for interaction in virtual reality. Adv. Sci. Technol. Res. J. **11**, 279–285 (2017)

Haghighi, A., Yang, Y., Li, L.: Dimensional performance of as-built assemblies in polyjet additive manufacturing process. In: ASME 2017 12th International Manufacturing Science and Engineering Conference. MSEC 2017 Collocated with JSME/ASME 2017 6th International Conference on Materials and Processing, vol. 2 (2017). https://doi.org/10.1115/MSEC2017-2983

Haidiezul, A.H.M., Aiman, A.F., Bakar, B.: Surface finish effects using coating method on 3D printing (FDM) parts. IOP Conf. Ser. Mater. Sci. Eng. **318**, 12065 (2018). https://doi.org/10.1088/1757-899X/318/1/012065

Leary, M., Babaee, M., Brandt, M., Subic, A.: Feasible build orientations for self-supporting fused deposition manufacture: a novel approach to space-filling tesselated geometries. Adv. Mater. Res. **633**, 148–168 (2013). https://doi.org/10.4028/WWW.SCIENTIFIC.NET/AMR.633.148

Singh, S., Sachdeva, A., Sharma, V.S.: Investigation of dimensional accuracy/mechanical properties of part produced by selective laser sintering. Int. J. Appl. Sci. Eng. Res. **10**, 59–68 (2013)

Sood, A.K., Ohdar, R.K., Mahapatra, S.S.: Grey Taguchi method for improving dimensional accuracy of FDM process. Mater. Des. **30**, 4243–4252 (2009)

Wang, C.C., Lin, T.W., Hu, S.S.: Optimizing the rapid prototyping process by integrating the Taguchi method with the Gray relational analysis. Rapid Prototyping J. **13**, 304–315 (2007). https://doi.org/10.1108/13552540710824814

Wong, K.V., Hernandez, A.: A review of additive manufacturing. ISRN Mech. Eng. **2012**, 1–10 (2012). https://doi.org/10.5402/2012/208760

Zharylkassyn, B., Perveen, A., Talamona, D.: Effect of process parameters and materials on the dimensional accuracy of FDM parts. Mater. Today Proc. **44**, 1307–1311 (2021). https://doi.org/10.1016/j.matpr.2020.11.332

Thermal Study for Optimization of the Thermomechanical Welding Butt Fusion Process of the PEHD Tubes (PE100)

Awadi Walid[1,2](✉) and Zidi Mondher[1]

[1] Laboratoire de Génie Mécanique LGM, Ecole Nationale d'Ingénieurs de Monastir, Avenue Ibn El Jazzar, 5019 Monastir, Tunisia
awadi_walid@yahoo.fr, mondher.zidi@enim.rnu.tn

[2] Institut Supérieur des Etudes Technologiques de Jendouba, Campus Universitaire Jendouba Nord, 8189 Jendouba, Tunisia

Abstract. Today, high-density polyethylene (HDPE) is the frequently used polymer in the world for public water and gas distribution, and also for the disposal of wastewater. The welding butt fusion technique of polyethylene tubes (HDPE) is that most chosen in the thermoplastic industry given its efficiency and advantages over other processes and through which the mechanical behavior of the welding can approach that of the original materials. We focus our study in this document on the study of distribution and the effect of temperature at the welding juncture, modified welding parameters: heating temperature, heating time and welding force (for fixing the other welding parameters). The issue of this article is the control of the temperature and the thickness of the molten zone allows a partial improvement in a technical origin of the welding joint's quality.

Keywords: Height density polyethylene (HDPE – PE100) · heating temperature · welding strength · thermomechanical butt welding

1 Introduction

Today, a very wide variety of products designed and manufactured in polymer. For high-density polyethylene (HDPE), the substantial gains in costs, working times, its lightness which facilitates handling, its good resistance to corrosion and its impact resistance are at the origin of recourse by manufacturers to its use. HDPE pipes are currently intended for public gas and water transport works. Given the long distances of the installations, it is necessary to ensure the assembly between the sections of the tubes, to do this there are enough assembly techniques. The use of butt welding technique is adopted as joining process of high density polyethylene (HDPE) pipes for different diameters. Welding parameters such as heating temperature, heating time, welding force (pressure), cooling time are the determining factors of the quality of the weld joint. Research work has led to the fact that the mechanical and microstructural characteristics in a welded joint of polyethylene pipes differ from one layer to another (inner layer – middle layer and top

T. Mabrouki et al. (Eds.): AIAM 2023, LNME, pp. 114–124, 2024.
https://doi.org/10.1007/978-3-031-47784-3_15

layer) [2, 5]. In the inner layer, the modulus of elasticity [3], the yield strength and the crystallinity rate are higher than those of the other layers, especially the outer layer. With the variation of the rate of crystallinity there was variation of the hardness from one layer to another [4]. Some mechanical properties of HDPE polyethylene in the weld joint remain almost the same as those of the original material, such as yield strength and elongation at break [11]. While other mechanical properties such as toughness and fatigue cracking of the welded material was lower than that of the non-welded pipe [12–14]. The operation of heating the surfaces of the tubes to the molten state causes the increase of the thermal movement of the chains of polymer (polyethylene) in the fusion layer. Upon application of axial pressure to the two surfaces of the pipes, the molecular chains entangled in the fusion layer of one surface unravel under the mutual effects of thermal movement and pressure, and penetrate into the space void in the fusion layer of the other surface and become entangled with the polymer chains, then in the welded joints after cooling [9, 10]. In the welding zone, the molten HDPE undergoes deformation, molecular interpenetration and recrystallization, this leads to the formation of a new structure under stress imposed by the welding process. During the recrystallization phase applied stresses as well as thermally induced stresses are present at both sides of the weld bead. Hongbin Dai and Jun Peng [8] proved that the distance X between the lines of the dog bone shape (X shape) of the welding area decreases with the increase of welding pressure. Polymer chains in welded joints have a radial orientation, the number of radially oriented chains increases with increasing pressure, while the axial polymer chains decrease and the size and density of the cracks decrease. The researchers explained by this, the decrease of some tensile mechanical properties of the welded joint. Therefore, it is necessary to master the thermal study to choose the appropriate force to be applied, the aim of which is to minimize the number of radial chains and to keep the maximum possible of the axial chains to have better mechanical characteristics. Welding defects may exist as a result of poor choices of welding parameters, among which we cite the excess pressure applied instantly which causes the rejection of all the molten material to the outside and only the non-molten material remains in contact and this leads to the defect of lack of fusion [7]. This paper is the subject of an experimental study of thermomechanical butt welding with imposed and controlling force, focusing on the study of the distribution and the effect of temperature in the tube during the different phases of the welding cycle. Welding. As well as a modeling and identification of the different parameters made with Matlab.

2 Description of the Experimental

The works presented are carried out on sections of PE100 tubes with a diameter of 75 mm, a thickness of 4.5 mm, SDR17, PN10. Thermomechanical butt fusion welding is carried out on a VIRAX D-40-160 mm machine Fig. 1. In this study, the tests were performed automatically (imposed force). Improvements have been made to this machine Fig. 2, an electric cylinder has been added to ensure automatic movement and control its mobile spindle. An application was developed on LabVIEW to operate the system. The latter consists of an electric actuator, a MYRIO card, a power supply card and two force and position sensors. In this way, the data of the various tests are checked, measured and recorded (Figs. 3, 4, 5 and 6).

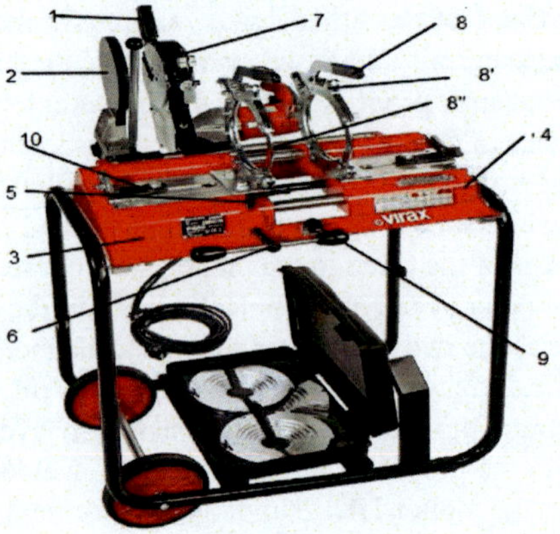

1 : Milling Equipment
2 : thermal elements
3 : fixed machine base
4 : mobile machine stand
5 : lock lever
6 : handlebar
7 : safety gate
8 : basic clamping jaw model
8' : regulating nut
8'' : clamping screw
9 : button mushroom
10 : drawstring

Fig. 1. VIRAX D40-160 heat welding machine

Fig. 2. VIRAX D40-160 heat welding machine with automation and servo modification

Fig. 3. Installation of the two ends of the tubes

Fig. 4. Installation of heating mirror (equalization heating)

Fig. 5. Mirror removal phase

Fig. 6. Welding & cooling phase

The tests were carried out in the ambient conditions of the mechanical workshop of the Higher Institute of Technological Studies of Jendouba (ISETJ), the measurement of the real conditions is made with a thermo-hygrometer brand CHAUVIN ARNOUX (Fig. 7), Humidity is 64.5%, Ambient temperature 22.5 °C. Similarly, during the tests, infrared thermal photos are recorded using an infrared thermal camera IR C.A 1886 (Fig. 8).

Fig. 7. Thermo-hygrometer CA1246

(a) (b) (c)

Fig. 8. (a), (b) and (c) Sampling infrared thermal images for a C.A 1886 IR thermal camera

Figures 9 and 10 are infrared thermal images taken by the camera in order to record the temperature variation in the solder joint at the different phases.

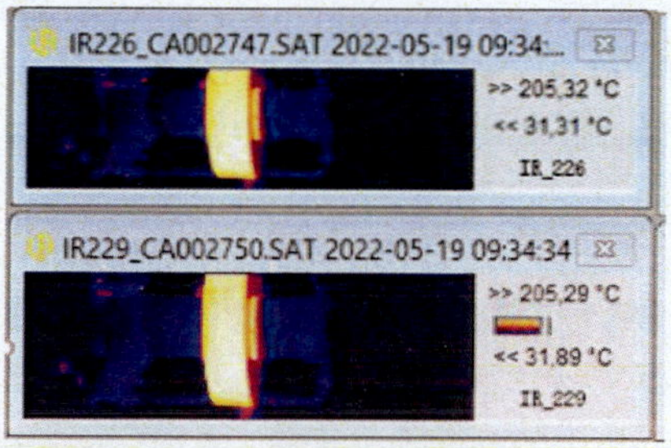

Fig. 9. Infrared thermal images of the equalization and heating phase

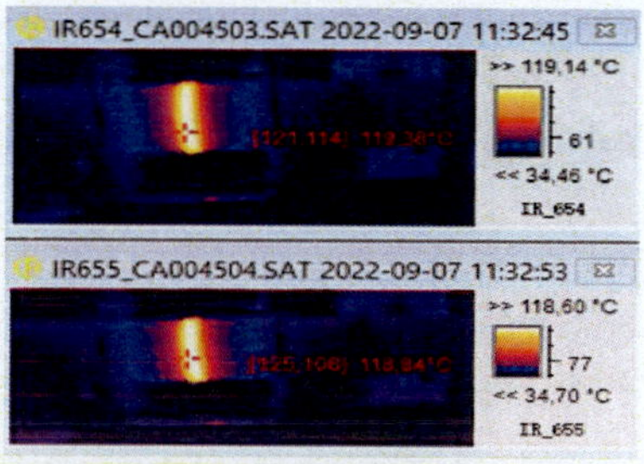

Fig. 10. Infrared thermal images of the welding and cooling phase

3 Results and Interpretation

3.1 Study of the Temperature During the Equalization and Heating Phases

During these two phases, the two sections of the tubes are subjected to thermal flows by the three modes (conduction, convection and radiation). The propagation of the temperature takes place radially and also axially.

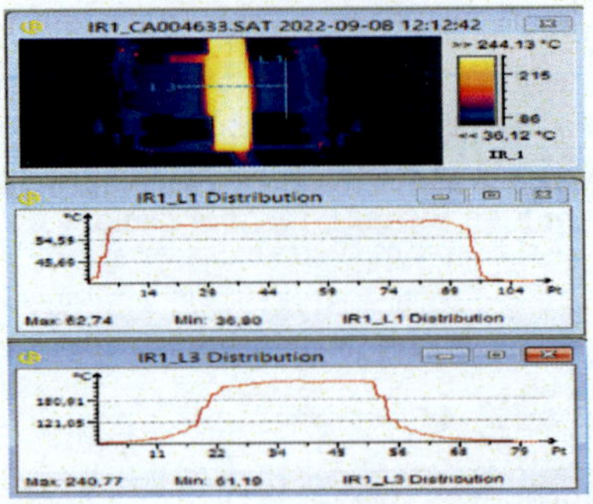

Fig. 11. The distribution of the temperature according to L1 is the same for constant x, the distribution according to L3 shows that the temperature depends on x

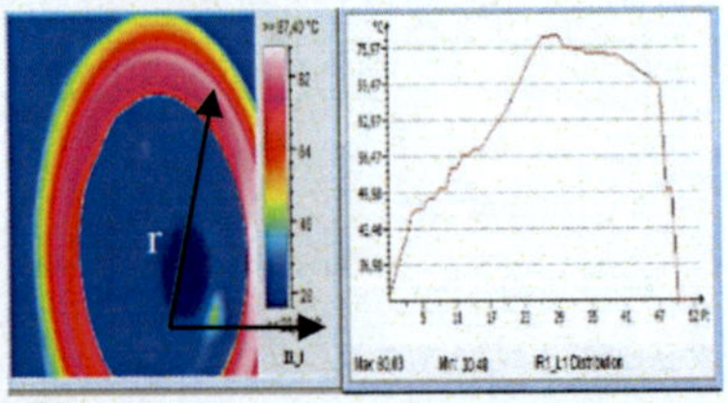

Fig. 12. Circumferential infrared thermal image shows that the temperature distribution depends on the radius and not on the angle

Figure 12 shows that the upper layer cools more quickly than the inner layer and consequently the middle layer and this confirms the variation of the mechanical properties and the rate of crystallinity from one layer to another, Figs. 11 and 12 clearly show that the temperature at the level of the tubes depends on time, on the distance X, the radius r and does not depend on the angle θ, so T (X, r, t), in this paper, we assume that the variation along the radius is small during heating since the thickness is small, and the width of the molten layer is small, we will therefore have T (X, t). During the tests we carried out, we recorded the same approach and the same temperature distribution as Mr. Olivier Hehn [1]. Figure 13 shows the evolution and the distribution of the temperature at different points which are at different distances during all the phases (equalization, heating, welding and cooling). The figure also shows that there are places where the temperature remains below the melting temperature.

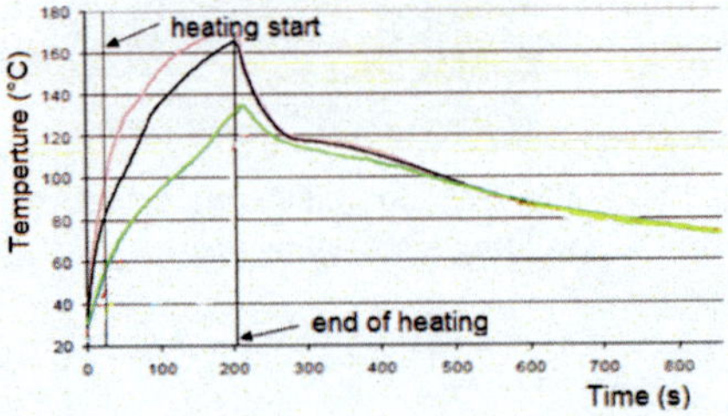

Fig. 13. The shape of the evolution of the temperature at the center of the wall of the tube as a function of time and of the initial distance to the welding plane

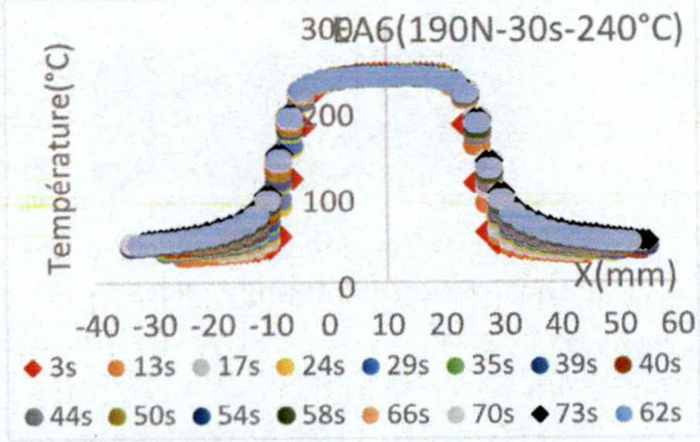

Fig. 14. Curves of the evolution of the temperature as a function of the distance for different instants during the phase of equalization and heating.

We present here in Fig. 14, an example of the EA6 test, the evolution of the temperature according to the distance X and the time t during the two phases of equalization and heating. The thickness of the heating plate equal to 20 mm (in this figure from X = 0 to 20 mm).

The mathematical model which describes the shape of the temperature variation as a function of the distance X (which separates a point of the tube from the heating plate) and the time t is determined using the MATLAB software and it is in the form:

$$T^{\circ}(x, t) = A.\exp\left[-k1.x^{n1} + B.\exp\left(-k2.x^{n2}\right) + C.(t)^{n3}\right] + Ta \quad (1)$$

A, k1, n1, B, k2, n2, c, n3 et Ta: are parameters to be identified.

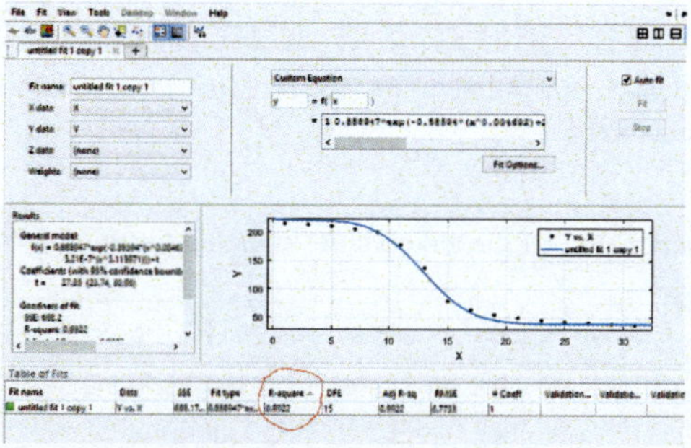

Fig. 15. Determination of the interpolation equation of the curve by Matlab

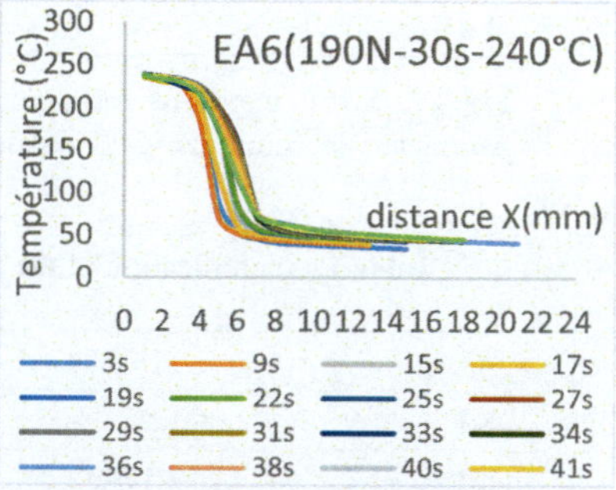

Fig. 16. Experimental curves of temperature as a function of distance at different times of a single section

In Fig. 15, the equation sticks with the curve has a quadratic error adj. R-sq. = 99.22%, this makes it possible to trust this model, to identify the parameters, we used the Excel solver which allows the calculation of the parameters by minimizing the quadratic error at the level of each point.

Figures 16 and 17 clearly show that the experimental curves and those calculated from Eq. (1) match for the identified values of the parameters in Table 1.

For this test EA6, at x = 0, T (0, t) = TM = 240 °C and the temperature is equal to Ta = 43 °C far from the mirror, at the end of the heating phase, we have the temperature is equal to 130 °C at a distance x equal 5.9 mm. By varying the test parameters, the thickness of the molten layer (Xfusion) varies, Table 2 illustrates this variation. The relationship between the thickness of the molten layer and the test parameters was modeled by the simple tool Minitab. The relation is linear (Eq. 2), it is verified by a midpoint.

Factorial regression: Xfusion as a function of Fs; t; T° (Descending elimination of terms); Candidate terms are: Fs; t; temp; Fs*t; Fs*T°; t*T°; Fs*t*T°.

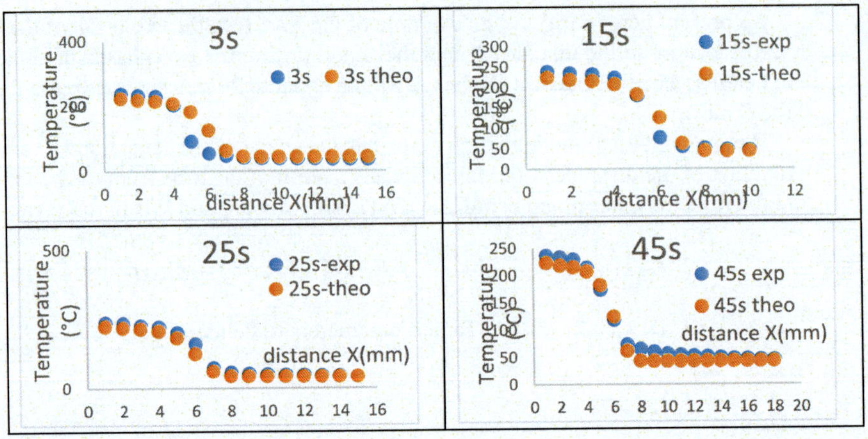

Fig. 17. Some experimental-theoretical figures drawn on Excel after identification of the parameters by minimizing the quadratic error

Table 1. Identification of the parameters of the temperature equation for the EA6 test

A	22.0043	B	14.758
K1	1.9E−6	K2	1.9494
n 1	7.17492	n2	0.0094
C	0.001	n3	0.0469

Table 2. Xfusion distance between heating plate and the farthest fusion point at 130 °C of the various tests

	Fs(N)	T(s)	T°(°C)	EA Xfusion(mm)
EA1	110	30	200	5,2
EA2	190	30	200	6,1
EA3	110	70	200	5
EA4	190	70	200	7,35
EA5	110	30	240	5,1
EA6	190	30	240	6.03
EA7	110	70	240	7,75
EA8	190	70	240	8,5
EA9	150	50	220	6,4

S	R carré	R carré (ajust)	R carré (prév)
0,529729	93,24%	84,22%	51,91%

Regression equation in uncoded units:

$$X_{fusion} = 11 + 0.01541.F_s - 0.241.t - 0.0403.T^\circ + 0.001272.t.T^\circ \quad (2)$$

Validation by midpoint EA9: Theoretical Xfusion = 6.3875 mm; Experimental Xfusion = 6.4 mm; Error = 0.0125 mm which represents 0.195%

3.2 Study of the Temperature During the Welding and Cooling Phases

During the welding and cooling operation, the temperature decreases in three stages Fig. 19: Stage A: decreases linearly up to 120 °C/Stage B: remains constant at 120 °C for a few seconds/And Stage C: decreases according to a power law.

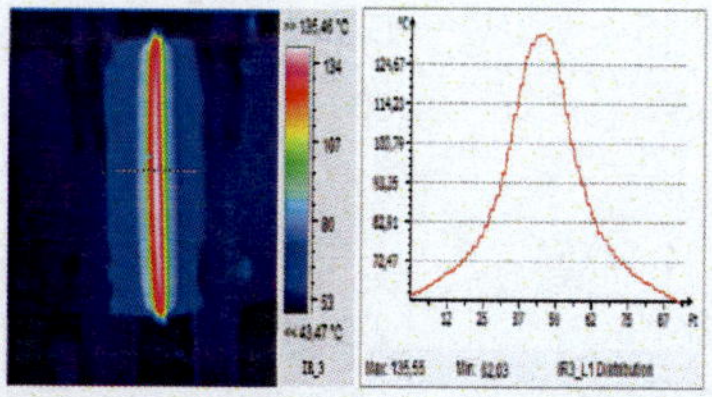

Fig. 18. Longitudinal infrared thermal image during the welding-cooling phase

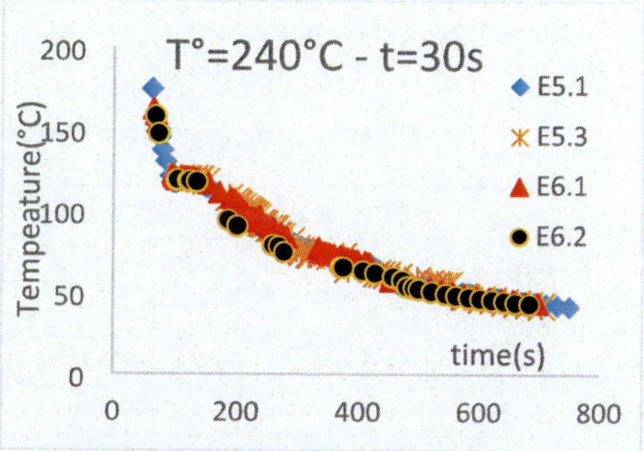

Fig. 19. Temperature decrease curve of tests EA5 and EA6 during the welding and cooling phases

During the cooling phase the temperature, as a function of x, keeps the same pace as that of the heating phase (Fig. 18). Figure 19 shows the decrease in temperature as a function of time at a point in the welding zone during the cooling phase (Fig. 20 and Table 3).

$$\text{Stage A}: \ T^\circ = a.t + b \quad (3)$$

Stage B: T° constant 120 °C (temperature of crystallisation)

$$\text{Stage C}: \ T^\circ = k.t^n \quad (4)$$

$$a = -0.034.T^\circ - 0.0138.t + 7.0733 \quad (5)$$

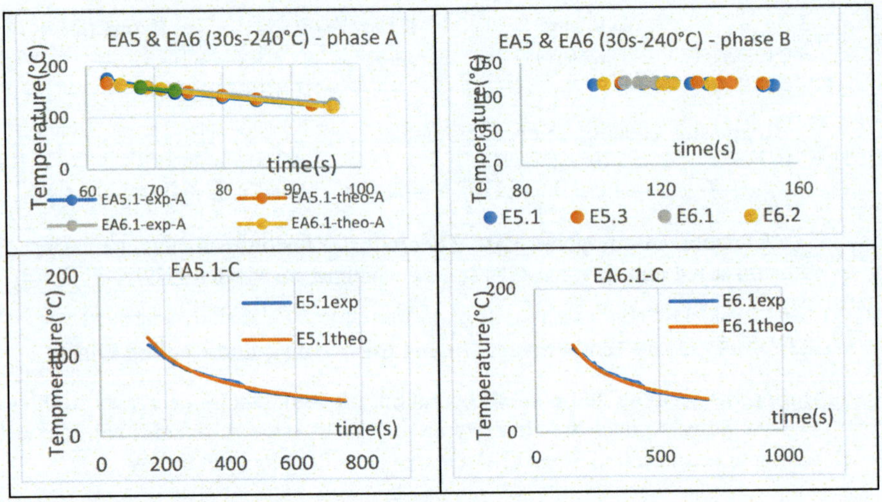

Fig. 20. Experimental and theoretical temperature reduction curves for tests EA5 and EA6 during the welding and cooling phases

Table 3. Summary for the identification of the coefficients a, b, k and n:

	EA1 & EA2	EA3 & EA4	EA5 & EA6	EA7 & EA8	EA9
t (s)	30	70	30	70	50
T° (°C)	200	200	240	240	220
a	−0,15	−0,70454	−1,51154	−1,14558	−1,1185
b	157	87,1939	261,5654	285,3223	254,169
k	1387,69	3797,75	3405,99	1916,683	2233,45
n	−0,5155	−0,66540	−0,66170	−0,53402	−0,5732

$$b = 0.05848.T^{\circ}.t + 0.8598.T^{\circ} - 13.4415.t + 37.445 \tag{6}$$

$$k = 2.437.T^{\circ}.t + 123.568.T^{\circ} + 547.667.t - 25133.56 \tag{7}$$

$$n = 2.6E - 5.T^{\circ}.t - 0.00441.T^{\circ} - 0.008957.t + 0.48 \tag{8}$$

Table 4 shows the identification of the coefficients of the theoretical models of the temperature variation during the cooling of the weld joint as a function of the welding parameters. All the tests have a stage B (plateau) practically at the same temperature 120 °C this means that the stage B is a property of the material and does not depend on the welding conditions, it is the crystallization stage.

Table 4. Verification with the midpoint (the EA9 test)

	experimental	calculation	% error
a	−1.118	−1.108	0.95%
b	254.169	197.767	22.19%
k	2233.4	2627	14.94%
n	−0.57329	−0.65278	12.17%

4 Conclusion

The study of the temperature during the different phases of the welding process makes it possible to control the welding process by prediction of the thickness of the molten layer depending on the welding conditions. It has been shown that the thickness of the molten material depends linearly on the welding parameters. This gives for us the possibilities to choose the suitable welding force which allows us to weld without achieving contact be-tween the non-melted materials of the two sections. The contact between the two non-melted ends is among the factors of welding defect (lack of fusion at the joint). The choice of a necessary force which ensures the welding makes it possible to minimize the density of the molecular chains which change their orientations from axial to radial.

Acknowledgements. This work is partially supported by AIAM'2023 conference, I'm really gratefully acknowledge the helpful comments and suggestions of the AIAM'2023 reviewers, which have improved my presentation.

References

1. Olivier Hehn, M.: Analyse expérimentale et simulation thermomécanique du soudage bout à bout de tubes de Polyéthylène. Thèse de Doctorat, Paris (2006)
2. Latifa, A.: Mémoire de Magister: comportement mécanique et rupture du HDPE sous environnements contrôlés, université de Badji Mokhtar – Annaba, Algérie (2007)
3. Niou, S., et al.: A method for mechanical property assessment across butt fusion welded polyethylene pipes. Int. J. Adv. Manuf. Technol. (2018) **97**, 543–561 (2018). Part of Springer
4. Moreno-Moreno, M., et al.: Mechanical and thermal properties of friction-stir welded joints of high density polyethylene using a non-rotational shoulder tool - Unfried-Silgado5, May 2018
5. Olesik, P., et al.: Structure and mechanical properties of high-density polyethylene composites reinforced with glassy carbon (2021)
6. Fascicule de brevet Europeen: Procede pour souder bout a bout deux pièces plastiques a code d'identification, au moyen d'une machine d'electrosoudage a commande automatique
7. Technical Letter Report Preliminary Assessment of NDE Methods on Inspection of HDPE Butt Fusion Piping Joints for Lack of Fusion - JCN N6398, Task 2D, S. L. Crawford/S. E. Cumblidge/S. R. Doctor/T. E. Hall/M. T. Anderson, May 2008
8. Dai, H., Peng, J.: Les effects des caractéristiques du joint soudé sur ses propriétés en soudage par fusion thermique HDPE. Mod. Phys. Lett. B, School of Materials Science and Engineering, Harbin University of Science and Technology **31**(15), 1750185 (2017)

9. Xie, C.F.: Sci. Technol. Innov. Herald **11**, 51 (2012)
10. Zhao, H.: China Plastics **25**, 48 (2011)
11. Ezrin, M.: Plastics Failure Guide—Cause and Prevention, Hanser, München (1996)
12. Chen, H., Scavuzzo, R.J., Srivatsan, T.S.: J. Mater. Eng. Perform. **6**, 473 (1997)
13. Marshall, G.P.: Advances in joining plastics and composites, TWI, 202 (1991)
14. Murch, M.G., Troughton, M.J.: A study of the applicability of the tensile, weld test for thick walled polyethylene pipe. TWI 757 (1993)

Additive Manufacturing and Investment Casting Comparison of Superalloys: Aerospace industry

B. Aouadi[1,5], S. Ghannem[1], B. Ben Fathallah[2,3](✉), and M. A. Yallese[4]

[1] Applied Mechanics and Engineering Laboratory (LR-11-ES19), University of Tunis El Manar, ENIT, BP 37, Le Belvédère, 1002 Tunis, Tunisia
abadreddine@pursuitaero.com, saoussen.ghannem@enit.utm.tn
[2] Laboratoire de Mécanique, Matériaux Et Procédés (LMMP), LR 99ES05, École Nationale Supérieure d'Ingénieurs de Tunis (ENSIT), Université de Tunis (UT), 5 Avenue Taha Hussien, B.P 56, Bab Menara, 1008 Tunis, Tunisia
brahim.benfathallah@enit.utm.tn
[3] ENIT, University of Tunis El Manar, BP 37, Le Belvédère, 1002 Tunis, Tunisia
[4] Mechanics and Structures Research Laboratory (LMS), May 8Th 1945 University, P.O. Box 401, 24000 Guelma, Algeria
yallese.mohamedathmane@univ-guelma.dz
[5] EUROCAST (Pursuit Aerospace), GP1, Route de Sousse KM7, 2033 Megrine, Tunis, Tunisia

Abstract. Investment casting is one of the oldest technologies used in the world. Recently, metal additive manufacturing (AM) has been evolving rapidly. Indeed, AM is defined as a process of shaping a part by adding a layer-over-layer. However, AM is limited to the production of parts in small and medium series. This paper consists of a review of the AM versus the Investment casting technology of super alloys such as Nickel and Cobalt based alloys. Moreover, it shows a comparison between processes complexity, advantages, disadvantages, product quality, production cycle time and production costs.

Keywords: Investment Casting · Additive Manufacturing · AM Processes · 3D printing · superalloys · aerospace industry

1 Introduction

In recent years, AM witnessed a very quick and high evolution in term of technology and applications fields. Additive manufacturing (AM) is based on the usage 3D CAD model, sliced into 2D sections, which allows to the part to be printed layer-over-layer (Careri et al. 2021). AM is used for many types of materials. Most common AM methods, used for metal printing and that were tested by Eurocast, we can list DED (Direct Energy Deposition), SLS (Selective Laser Sintering) and SLM (Selective Laser Melting) (Careri et al. 2021).

Investment Casting is one of the most ancient manufacturing methods in the world. It is one of the main technologies used by aerospace industry to produce engines high temperature alloy parts. Relatively simple process with very high accuracy and net shape

T. Mabrouki et al. (Eds.): AIAM 2023, LNME, pp. 125–131, 2024.
https://doi.org/10.1007/978-3-031-47784-3_16

results, the basic principles of investment casting technology did not change thru the years. AM and Investment Casting are completely different technologies with many similarities that can be witnessed and tested on the parts produced by these two technologies such as mechanical proprieties, internal and external complex shapes (Vevers et al. 2018; Yueling et al. 2022).

The main disadvantages of the investment casting technology are: long lead time and high cost to design and produce a pattern wax tooling. Additive manufacturing does not have these limitations. With AM, we can start producing parts immediately from a 3D file. CAD model modification is easy and quick comparing to wax pattern tool design and manufacturing (Vevers et al. 2018; Yueling et al. 2022).

The rapid development of additive manufacturing technology drove many searchers to compare this technique to the investment casting technique and to determine if it will be the near-future technology that will replace investment casting process.

2 Investment Casting

Investment casting or lost wax casting is particularly applicable to the foundry of materials with high melting point. This high melting point foundry requires a mold made with refractory material (ceramic).

This molding process applies to all types of parts, from small to large sizes and from a few grams to many kilograms pieces, with simple and complicated shapes. Tolerances wide to tighter tolerances and surface finishes roughness from 2 to 3 μm. For Investment casting process, wax model is injected, assembled into one tree and then dipped into multiple ceramic slurries mixtures to create a shell that is called mold. Using an autoclave machine, the wax inside the mold is melted out. Pre-heating mold, to burn wax residue and prevent thermal shock when we pour the molten metal inside the mold (Kanyo et al. 2020). After solidification, the outside shell is knocked out and the metal parts are cut off from the mold runners. Investment casting is used in many fields mainly we list aerospace, medical, industrial gas turbine, etc.

Investment Casting Process as shown in Fig. 1.

3 Additive Manufacturing

Thru the years, AM become one of the most popular technology the last decades. Very useful when we are seeking for a rapid prototyping and for a soft material printing (Ghoussoub et al. 2023). Additive Manufacturing application is widely developed, and the scientists focus moved from soft materials to harder metal parts (Nickel and cobalt based alloys) that can be used in aerospace, energy generation and many other industries. Additive Manufacturing technology can be summarized in five steps. Start with development of a 3D CAD model. Then convert it to a standard format recognized by the Additive Manufacturing machine software. Once file is transferred to the AM machine, the part is then treated as end-users wants rescaling, changing position or orientation, etc. After completing the model treatment, the part manufacturing starts, and it will be built layer-over-layer by the Additive Manufacturing machine. Finally, the model will be cleaned and finished as needed.

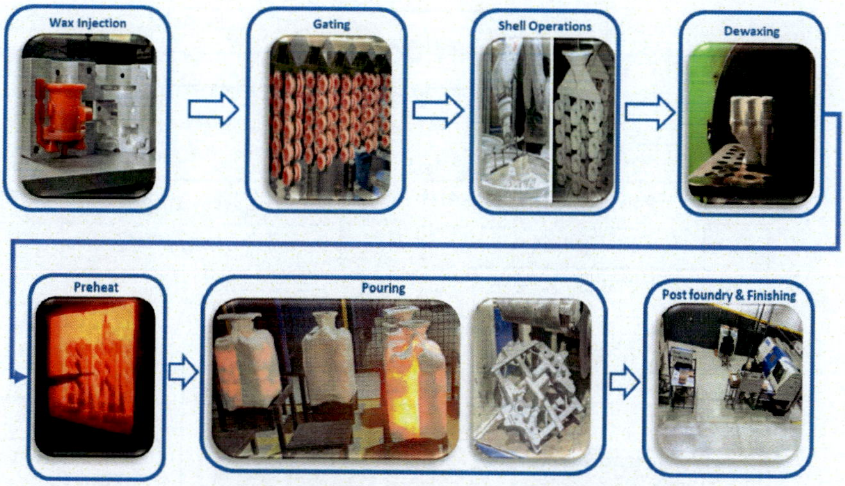

Fig. 1. Investment Casting Process 'Eurocast Tunisia'

3.1 AM - Processes

Organigram below (Fig. 2) shows all AM methods (Zhu et al. 2017; Prakash et al. 2018) and the highlighted one, they were used by Eurocast Tunisia to produce prototypes and compare them to investment casting process.

Table 1 summarize the comparison between the three methods.

3.2 AM - High Temperature Applications Alloys

AM technology did not well perform with high temperature alloys especially the ones used to manufacture engines high section's components (Bandyopadhyay et al. 2019). High temperature applications, and mainly the aerospace sector, they use Nickel based alloys such as Rene 108, INCO 718 and Mar M247 for these categories of components. However, the previously stated alloys have a poor weldability due to cracks generated during the welding process. Therefore, investment casting is the best way to pour this type of material. (Careri et al. 2021).

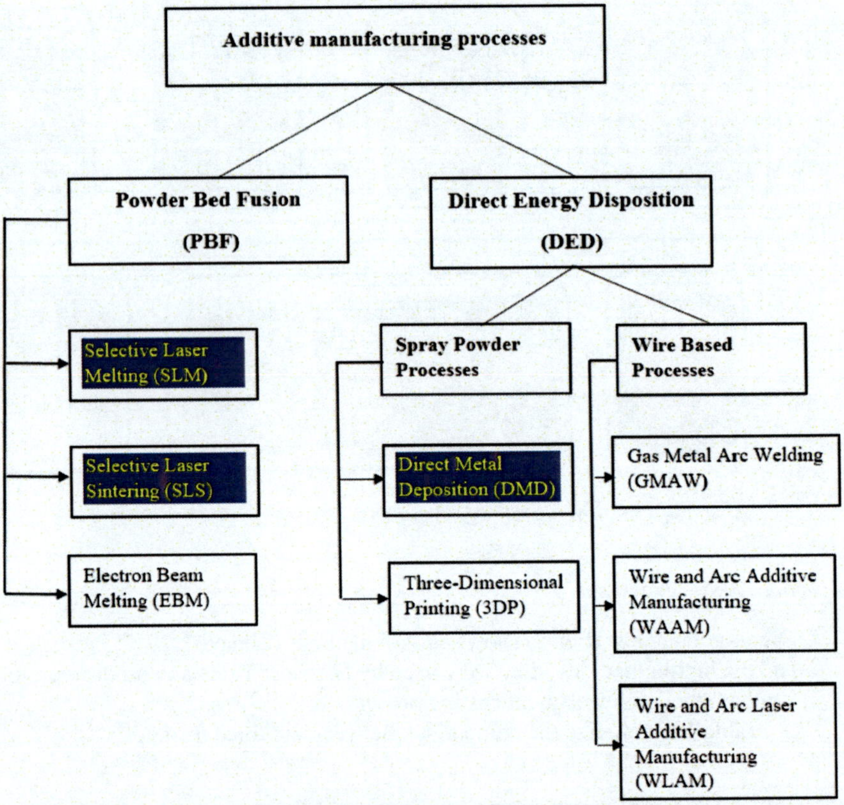

Fig. 2. Classification of AM processes

4 Additive Manufacturing vs Investment Casting

Some literature research studied and made comparison of mechanical properties (hardness), surface roughness, tensile strength and microstructure of samples produced from AM to samples produced with investment casting process (Table 2). The benefits of investment casting and additive manufacturing are presented in Table 3.

Eurocast (Pursuit Aerospace) an investment casting company manufacturing high temperature alloys components for aerospace engines compared some properties of AM and investment casting parts and below some results.

Table 1. SLM -SLS and DED Comparison (Shahryar Mooraj et al. 2021; Graybill et al. 2018).

SLM (Selective Laser Melting)	SLS (Selective Laser Sintering)	DMD (Direct Metal Deposition)
Melt selectively powder layers to create a net shape product using high power laser source. Advantages: Wide range of materials, high-performance with complex detailed shapes. Disadvantages: Rough surface finish that require a lot of post-processing. Very expensive and slow manufacturing process.	Powder is sintered the utilization of a Co2 laser beam. The part is created layer over layer and the process is driven by the movement of piston which is identical to the layer thickness. Advantages: Wide range of materials and quickly build of complex parts. Disadvantages: Slightly rough surface, and powder used can be potentially hazardous.	Powder jet technique that can be used for repair, hard facing and reconfiguration of forging dies. The technique has also been used to add features. Advantages: Very useful for functional parts repair. Low-cost process. Accurate and near-net shape for small parts. Disadvantages: Low resolution and poor surface finish. Limited to certain geometry.
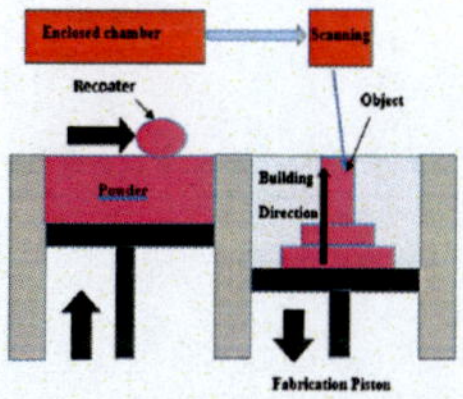	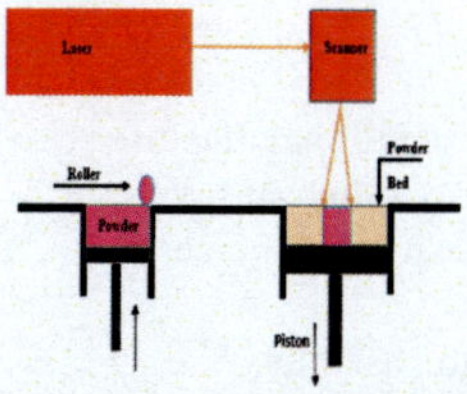	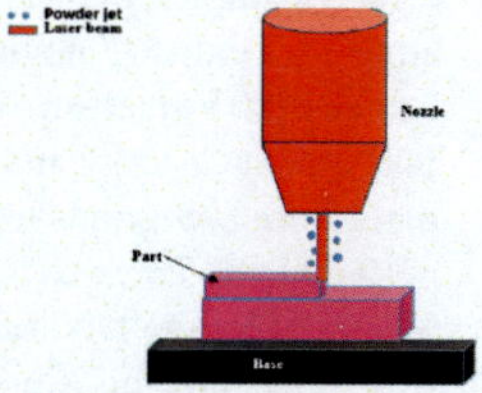

Table 2. AM and Investment casting comparison.

Properties	AM	Investment Casting	Comparison
Hardness	About the same for the entire part, regardless the thicknesses variation of features	High for thin walls and low for thicker sections	For AM, the hardness values are not depending on the features thickness
Surface roughness	80 μm	60 μm	AM surface quality is better than casting technology
Production time (h)	Very short	Too long	Investment casting is more suitable for series production
Production costs	Excessively high for small batches	Very competitive for medium-Big size batches	AM technology is more suitable for small batches Investment casting tooling design and manufacturing cost is excessively high

Table 3. Benefits of investment Casting and AM.

Investment Casting	AM
Accurate and tight tolerances Tight tolerance and accuracy are one of the most important capabilities of Investment casting **Revert Material Recycling** Revert generated from casting molds can be refined and reused for another investment casting pouring **Net shapes and surface texture Quality** Net shapes, accurate features reproduction and surface texture quality are the strengths of investment casting process **Simple to Complex Shapes** Investment casting is used to produce simple to very complex parts shapes and designs for almost all industry fields **Large Size Variation** Investment casting can be used to manufacture parts from few grams to many kilograms with perfect compliance to the desired tolerances **Material Usage Diversity** Investment casting is used to cast a large variety of alloys, such as and not limited to, Stainless Steel, High temperature alloys, nickel, Titanium, cobalt, aluminum, and many others	**Flexible Design** Adequate for more complex designs comparing to traditional casting processes **Rapid Prototyping** Parts could be produced in few hours with complex shapes. Very competitive comparing to traditional processes Print on Demand Very economic method in term of space and resources consuming No tooling needed; the 3D design files are all stored in a virtual library Easy and non-expensive design upgrade of 3D models

5 Conclusion

Additive manufacturing is the creation of 3D parts from a CAD file. AM is, currently, widely used because it offers a new method to quick manufacture a complex shaped part. The fast evolution of additive manufacturing technology is due to its advantages comparing to the conventional manufacturing techniques.

Additive manufacturing can be used in all industrial sectors, and more and more is replacing conventional methods of casting especially for prototyping manufacturing.

Research' test results have proved that AM is the near future substitution of the casting parts. Recent studies showed that mechanical properties of parts produced from AM are similar and even better, in some cases, than parts produced from investment casting process. Production costs and lead time of large batches, remain the major disadvantage of Additive Manufacturing. Small batches (1 to 5 pcs), AM is very competitive comparing to investment casting process.

Acknowledgements. This study was elaborated in a collaboration between three laboratories LMMP – ENSIT University of Tunis - Tunisia, LMAI - ENIT - El Manar University - Tunisia and

LMS - University 8 May 1945 Guelma - Algeria. We sincerely express our thanks to the industrial "Pursuit Aerospace" EUROCAST, Tunisia: investment casting company specialized in complex aerospace machined components.

We gratefully acknowledge the valuables contribution and comments provided by our reviewers to improve the Quality of this presentation.

References

Bandyopadhyay, A., Shivaram, A., Isik, M., Avila, J.D., Dernell, W.S., Bose, S.: Additively manufactured calcium phosphate reinforced CoCrMo alloy: bio-tribological and biocompatibility evaluation for load-bearing implants. Addit. Manuf. **28**, 312–324 (2019). https://doi.org/10.1016/j.addma.2019.04.020

Careri, F., Imbrogno, S., Umbrello, D., Attallah, M.M., Outeiro, J., Batista, A.C.: Machining and heat treatment as post-processing strategies for Ni-superalloys structures fabricated using direct energy deposition. J. Manuf. Process. **61**, 236–244 (2021). https://doi.org/10.1016/j.jmapro.2020.11.024

Ghoussoub, J.N., Utada, S., Pedraza, F., Dick-Cleland, W.J., Tang, Y.T., Reed, R.C.: Alloy design for additive manufacturing: early-stage oxidation of nickel-based superalloys. Metallurgical Mater. Trans. A **54A**, 1721–1729 (2023). https://doi.org/10.1007/s11661-022-06860-6

Graybill, B., Li, M., Malawey, D., Ma, C., Alvarado-Orozco, J.-M., Martinez-Franco, E.: Additive Manufacturing of Nickel-Based Superalloys (2018). http://www.asme.org/about-asme/terms-of-use

Kanyo, J.E., Schafföner, S., Uwanyuze, R.S., Leary, K.S.: An overview of ceramic molds for investment casting of nickel superalloys. J. Eur. Ceramic Soc. **40**(15), 4955–4973 (2020). Elsevier Ltd. https://doi.org/10.1016/j.jeurceramsoc.2020.07.013

Prakash, K.S., Nancharaih, T., Rao, V.V.S.: Additive manufacturing techniques in manufacturing -an overview. Mater. Today Proc. **5**(2), 3873–3882 (2018). https://doi.org/10.1016/j.matpr.2017.11.642

Yueling, G., Qifei, H., et al.: Comparative study on wire-arc additive manufacturing and conventional casting of Al–Si alloys: porosity, microstructure and mechanical property. Acta Metallurgica Sinica (English Lett.) **2022**(35), 475–485 (2022). https://doi.org/10.1007/s40195-021-01314-1

Vevers, A., Kromanis, A., Gerins, E., Ozolins, J.: Additive manufacturing and casting technology comparison: mechanical properties, productivity and cost benchmark. Latv. J. Phys. Tech. Sci. **55**(2), 56–63 (2018). https://doi.org/10.2478/lpts-2018-0013

Mooraj, S., Qi, Z., et al.: 3D printing of metal-based materials for renewable energy applications. Nano Res. **14**(7), 2105–2132 (2021). https://doi.org/10.1007/s12274-020-3230-x

Zhu, C., et al.: 3D printed functional nanomaterials for electrochemical energy storage. Nano Today **15**, 107–120 (2017). https://doi.org/10.1016/j.nantod.2017.06.007

The Impact of the Adaptive Slicing Integration on the AM Cost

Ahmed Elayeb[1](✉), Farhat Zemzemi[1], Mehdi Tlija[2], and Borhen Louhichi[1]

[1] National Engineering School of Sousse, Mechanical Laboratory of Sousse, (LMS, ENISo), University of Sousse, 4023 Sousse, Tunisia
elayebahmed@gmail.com, borhen.louhichi@etsmtl.ca

[2] National Engineering School of Monastir, Mechanical Engineering Laboratory (LGM, ENIM), University of Monastir, 5019 Monastir, Tunisia

Abstract. The manufacturing industry has been revolutionized by additive manufacturing (AM) by allowing the creation of complex geometries that were once impossible to manufacture using traditional methods. However, the layer-by-layer approach of additive manufacturing can result in dimensional and geometric errors due to the staircase effect, particularly in parts with complex shapes. Hence, achieving a balance between the quality and the cost is one of the most AM challenges. In this context, Design for Additive Manufacturing (DFAM) and adaptive slicing have emerged as promising approaches to reduce these errors while establishing a balance between the AM quality and cost. This work focuses on the importance of the adaptive slicing in minimizing the AM time and cost during the DFAM phase. A method for adaptive slicing is proposed based on minimizing errors due to the staircase effect caused by shape curvatures in the part. The results demonstrate that the proposed method can in some cases significantly save more than 80% of building time and cost in order to obtain the same surface quality comparing to the standard slicing method. Moreover, this slicing technique also improves the parts dimensional and geometrical tolerances, hence, reduces errors and improves the overall quality. The presented approach has the potential to enhance the capabilities of additive manufacturing, allowing the creation of parts with higher accuracy and precision, and reducing the need for post-processing.

Keywords: Additive manufacturing · cost saving · Adaptive slicing · time saving · DFAM

1 Introduction

The Additive Manufacturing (AM), also known 3d printing, is a rapidly growing field in the manufacturing industry, allowing the production of complex geometries that were not possible with traditional manufacturing methods. However, the printed parts often suffer from dimensional and geometrical errors due to the layer-by-layer printing process, commonly referred to as the staircase effect. To overcome these issues, Design for Additive Manufacturing (DFAM) is an important phase in the process of preparing the file to be printed. Choosing the right printing orientation combined with adaptive slicing

T. Mabrouki et al. (Eds.): AIAM 2023, LNME, pp. 132–138, 2024.
https://doi.org/10.1007/978-3-031-47784-3_17

can significantly reduce dimensional and geometrical errors, leading to improved part quality. Printing time may also be an issue for additive manufacturing. Enhancing the quality of the parts often requires a compromise between printing time and quality. So, a compromise between printing time and quality tends to be necessary. This paper is divided into three main sections: the state of the art, the developed method, and the discussion and conclusion.

To reduce dimensional and geometrical errors in AM parts, several research investigations have concentrated on the optimization of printing orientation and adaptive slicing algorithms. The two primary categories of slicing depending on the rectangular or sloped edge of the portion, such as adaptive slicing, were suggested (Pandey et al. 2003) after realizing an overview of the slicing techniques. In the same vein, (Binatra et al. 2021) investigated the relationship between layer height and surface roughness and discovered that a layer height ranges between 0.05 mm and 0.25 mm A layer height ranging between 0.05 mm and 0.25 mm offers a compromise between printing time and surface roughness. The best outcomes are achieved at layer heights of 0.15 mm and 0.2 mm, while the lowest surface roughness is achieved at 0.05 mm. Working with adaptive slicing, (Byun et al. 2006) investigated the ideal construction orientation. The construction orientation was one of the manufacturing variables that had the biggest effect on the direct cost of the parts that were produced in their research. (Zhao et al. 2000) reduced the layer thickness during 'uniform slicing', also known as 'direct slicing', to minimize the staircase effect. This involves cutting the 3D part with the same slice thickness over the whole section. The printout time and cost could increase significantly as a result. In adaptive slicing, the layer thickness was adjusted to take into account the vertical surface curvatures of the solid model, to lessen the staircase effect, and to minimize the number of layers to speed up build time.

One of the challenges of this manufacturing process is the lack of optimization for dimensional and geometrical tolerances in printed parts. Unacceptable part quality results from this lack of support for dimensional and geometric tolerances. An effective, assured, and quick option is to modify the print code (g-code) in the adaptive slicing process to account for both dimensions and geometric tolerances during printing. It enables a decent part that satisfies the requirements set out by the engineers during the Design For Additive Manufacturing (DFAM) stage. (Diegel 2019), (Berni 2021) To address this problem, some studies proposed various approaches for dimensional and geometrical tolerancing for AM, such as taking into account various AM parameters (printing support, bed temperature, nozzle, or printing head). The part orientation and layer thickness (Solomon et al. 2020) (Elayeb et al. 2023) (Elayeb et al. 2022).

Another recommended solution to DFAM optimization is the adaptive slic-ing technique (Mao et al. 2019). Overall, the combination of printing orientation and adaptive slicing strategies has shown great potential in minimizing the dimensional and geometrical errors in AM parts. In this context, the authors of this paper are currently studying the impact of the adaptive slicing on the minimization of the dimensional errors. After the experimental validation through a CMM machine, the preliminary results show that the adaptive slicing can reduce the dimensional error from 0.52% to 0.08% of a feature's dimension. Furthermore, the potential of adaptive slicing and other advanced techniques has been demonstrated for reducing printing time and improving efficiency in additive manufacturing. However, further research is needed to optimize these strategies for specific geometries and materials to improve part quality and reduce production costs.

2 State of the Art

Adaptive slicing is an innovative technique that aims to improve the surface quality and reduce printing time in FDM. The conventional slicing technique involves dividing the 3D model into layers of equal thickness and then generating toolpath instructions for each layer. However, this method may lead to inefficient printing and poor surface quality when dealing with complex geometries or curved surfaces. In contrast, adaptive slicing modifies the layer thickness based on the geometry of the 3D model, reducing the number of layers where possible and increasing the layer thickness in less complex areas. The proposed method in this paper uses an algorithmic approach to implement the adaptive slicing technique, which allows for quick and easy generation of optimal G-code for FDM printing of complex parts. The use of MATLAB software provides a flexible platform for algorithm development, allowing for the customization of the algorithm to meet the specific needs of different printing applications. This scientific paper presents a method of adaptive slicing for Fused Deposition Modeling (FDM) parts using a MATLAB algorithm that takes an STL file as input and generates G-code of standard and adaptive slicing as outputs. The proposed algorithm follows a series of steps, starting with importing the CAD file in the STL format and selecting the printing orientation. Then, common printing parameters are specified, and the coordinates of

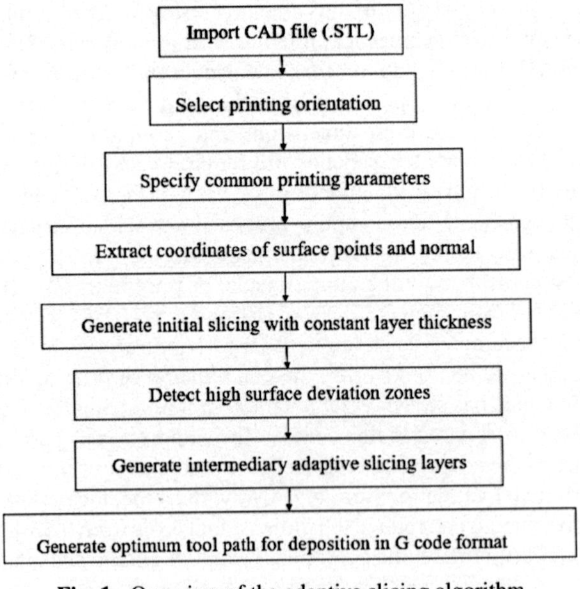

Fig. 1. Overview of the adaptive slicing algorithm

surface points and normal are extracted. The algorithm generates the standard slicing and detects high surface deviation zones, which are then used to generate the adaptive slicing. Finally, the algorithm generates the optimal toolpath and the G-code required for the FDM printing process. The main criteria for the application of the adaptive slicing in each layer is based on two conditions to control the surface deviation and the layer height. The First condition is to achieve the least surface deviation between the two successive layers. In our case, let S1 and S2 be, respectively, the two successive layers to control. The surface deviation should not exceed 0.45% of S1 otherwise a new layer is created in the mid distance of S1 and S2. The second condition concerns the layer height which should not be under the FDM machine capability. The minimum layer height is chosen to be 0.05 mm as it is commonly the minimum achievable layer height (Fig. 1).

3 Case Study

3.1 The Results

The designed parts are imported to the MATLAB code as STL files then the adaptive and standard slicing are performed as shown in Figs. 3 and 4. The AM quality is based on a user pre-chosen surface deviation of 0.0045 guaranteed by both of the standard and the adaptive slicing to achieve the same part quality (Fig. 2).

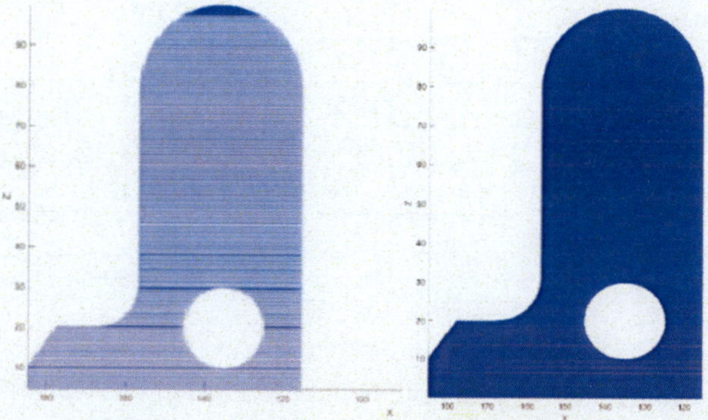

Fig. 2. Generated standard and adaptive slicing overview for part1

The generated G-code gives the following printing results shown in Table 1. The studied AM parameters and results for both pats are the filament weight used, the estimated building time, the layer height, the layer count, the time cost, the filament cost, the total cost, the time gain, the filament gain and the total cost gain.

3.2 The Discussion

The effectiveness of the proposed method is demonstrated through experiments on various test parts, and the results show that the adaptive slicing approach leads to improved

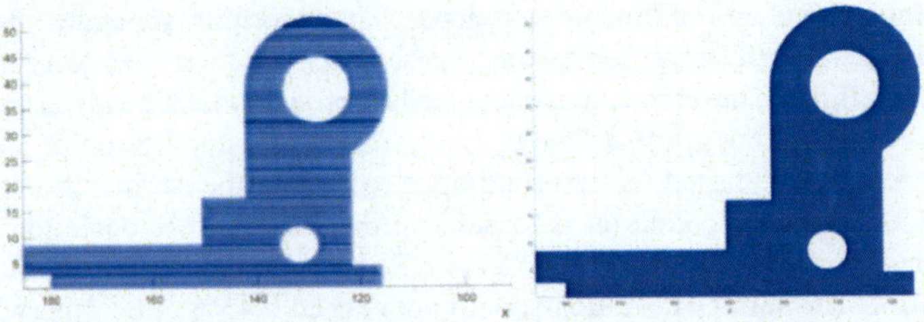

Fig. 3. Generated standard and adaptive slicing overview for part 2

Table 1. Dimensional accuracies of cylindrical features

The AM parameters and results	Part 1		Part 2	
	Part with standard slicing	Part with adaptive slicing	Part with standard slicing	Part with adaptive slicing
Filament weight used	25.41 g	25.02 g	10.15 g	10.13 g
Estimated building time	25:45:32	4:25:2	8:12:46	2:17:23
Layer height	0.05 mm	0.05 to 0.3 mm	0.05 mm	0.05 to 0.3 mm
Layer count	1998	371	1049	308
Time cost	25.76	4.42	8.21	2.29
Filament cost	1.27	1.25	0.51	0.51
Total cost	27.03	5.67	8.72	2.80
Time gain	82.85%		72.12%	
Filament gain	1.57%		0%	
Total cost gain	79.02%		67.89%	

surface quality and reduced printing time compared to standard slicing methods. The presented method provides a practical and efficient solution for generating G-code for FDM printing of complex parts with varying surface geometry.

The developed chart in Fig. 4. Demonstrates that the developed method has been executed on two different case studies in which different parameters are ameliorated such as printing time and filament cost. Also, it reduces building time by 82.85% and 72.12%. And total cost by 79.02 and 67.89% for part 1 and 2 respectively.

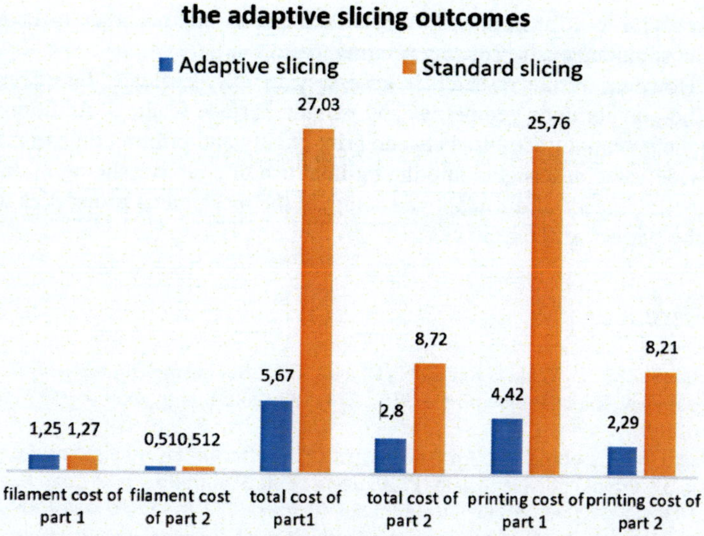

Fig. 4. Generated standard and adaptive slicing overview for part1

4 Conclusion

Additive Manufacturing (AM) has revolutionized the manufacturing industry by allowing the production of complex geometries that were previously impossible to achieve with traditional manufacturing methods. However, the layer-by-layer printing process of AM often leads to dimensional and geometrical errors, commonly known as the staircase effect. The Design for Additive Manufacturing (DFAM) phase plays a crucial role in preparing the file for printing and minimizing these errors.

The state of the art in this field shows that several research studies have focused on optimizing the slicing strategies to minimize dimensional and geometrical errors in AM parts. Different works proposed various optimization methods demonstrating significant improvements in the dimensional accuracy, mechanical properties, surface finish, and reduced printing time of AM parts. However, these methods have certain limitations, such as not being suitable for complex geometries, requiring extensive computational resources or training data, and being unsuitable for parts with varying surface features or irregular geometries.

In summary, the study proposes an adaptive slicing algorithm that utilizes MATLAB to generate G-code for both standard and adaptive slicing modes from an input STL file. The algorithm is shown to effectively reduce the printing time and minimize dimensional and geometrical errors compared to standard slicing. Furthermore, the results demonstrate a significant reduction in printing time when using adaptive slicing with a minimum layer height of 0.05 mm and a maximum layer height of 0.30 mm compared to the standard slicing method using a layer height of 0.05 mm to guarantee the same part's quality. These findings suggest that the proposed adaptive slicing algorithm can be a useful tool for improving the surface quality and accuracy of FDM printed parts while simultaneously reducing printing time.

However, future research is necessary to fully evaluate the effectiveness of this method on complex geometries and varying surface features. Additionally, investigating the potential trade-offs between print quality and printing time in more detail could provide valuable insights into the optimal use of adaptive slicing in different settings. Also, the impact of the adaptive slicing on the mechanical proprieties of the parts may be the subject of future works.

References

Mohan Pandey, P., Venkata Reddy, N., Dhande, S.G.: Slicing procedures in layered manufacturing: a review. Rapid Prototyping J. **9**(5), 274–288 (2003). https://doi.org/10.1108/13552540310502185

Bintara, R.D., Lubis, D.Z., Aji Pradana, Y.R.: The effect of layer height on the surface roughness in 3D printed polylactic acid (PLA) using FDM 3D printing. IOP Conf. Ser. Mater. Sci. Eng. **1034**(1), 012096 (2021). https://doi.org/10.1088/1757-899X/1034/1/012096

Byun, H.S., Lee, K.H.: Determination of optimal build direction in rapid prototyping with variable slicing. Int. J. Adv. Manuf. Technol. **28**(3–4), 307–313 (2006). https://doi.org/10.1007/s00170-004-2355-5

Zhao, Z., Laperrière, L.: Adaptive direct slicing of the solid model for rapid prototyping. Int. J. Prod. Res. **38**(1), 69–83 (2000). https://doi.org/10.1080/002075400189581

Diegel, O., Nordin, A., Motte, D.: DfAM strategic design considerations. In Walha,V., Jarraya, A., Djemal, F., Haddar, M., (eds.) Design and Modeling of Mechanical Systems – V, pp. 41-70. Springer (2019). https://doi.org/10.1007/978-981-13-8281-9_3

Berni, A., Borgianni, Y., Obi, M., et al.: Investigating perceived meanings and scopes of design for additive manufacturing. Proc. Design Soc. Int. Conf. Eng. Design **1**, 1937–1946 (2021). https://doi.org/10.1017/pds.2021.455

Solomon, I.J., Sevvel, P., Gunasekaran, J.: A review on the various processing parameters in FDM. Mater. Today Proc. **37**, 509–514 (2020). https://doi.org/10.1016/j.matpr.2020.05.484

Elayeb, A., Eltaief, A., Korbi, A., et al.: Printing orientation selection based on the dimensional errors modelling in additive manufacturing. In Walha, L., Jarraya, A., Djemal, F., Haddar, M., (eds.), Design and Modeling of Mechanical Systems - V (pp. 195-205). Springer, Cham (2023).https://doi.org/10.1007/978-3-031-14615-2_23

Elayeb, A., Korbi, A., Tlija, M., Louhichi, B.: Optimal part orientation in the additive manufacturing process based on tolerancing. In Bouraoui, T., Benameur, T., Mezlini, S., et al. (eds.) Advances in Mechanical Engineering and Mechanics II, pp. 217-225. Springer,Cham (2022). https://doi.org/10.1007/978-3-030-86446-0

Mao, H., Kwok, T.H., Chen, Y., Wang, C.C.L.: Adaptive slicing based on efficient profile analysis. Comput. Aided Des. **107**, 89–101 (2019). https://doi.org/10.1016/j.cad.2018.09.006

In-Situ Monitoring of Selective Laser Melted Ti–6Al–4V Parts Using Eddy Current Testing and Machine Learning

Haifa Sallem[1(✉)], Hatem Ghorbel[2], Edouard Goffinet[2], Adeline Cinna[1], Jean Pralong[1], Jonatan Wicht[3], and Bernard Revaz[3]

[1] Powder Technology and Advanced Materials, HEI-VS, HES-SO University of Applied Sciences and Arts Western Switzerland, Rue de l'Industrie 23, 1950 Sion, Switzerland
haifa.sallem@hevs.ch

[2] Data Analytics Group, HE-ARC Ingénierie, HES-SO University of Applied Sciences and Arts Western Switzerland, Rue de la Serre 7, CH-2610 Saint-Imier, Switzerland
hatem.ghorbel@he-arc.ch

[3] AMiquam SA, 2 Route Cité Ouest, 1196 Gland, Switzerland

Abstract. Metal laser powder bed fusion (L-BPF) technology is one of the most common and evolved additive manufacturing technologies to fabricate metal components. However, the control of defects generated during the SLM process remains an essential technological challenge for its implementation in production lines. In this work, based on the combination of eddy current testing (ECT) and machine learning (ML) approach, we propose a methodology allowing the in-situ monitoring of LPBF process porosity defects of Ti-6AL-4V components. The present empirical approach is achieved by setting up trained AI algorithms for the in-situ detection of porosity defects generated during the part fabrication. The algorithms are fed with data collected layer by layer using a specific experimental set up composed of an ECT system mounted on the machine recoater of the SLM machine. Comparison between predicted and experimental outcomes shows the effectiveness of the proposed framework which allows the prediction of porosity defects layer by layer with a mean absolute error (MAE) of 0.1% for CNN2D algorithm and 0.11% for LSTM one. The framework developed in this study can be effectively applied to quality control in additive manufacturing.

Keywords: Selective laser melting · Porosity · Defects · Monitoring · Eddy Current · Artificial Intelligence · Machine learning

1 Introduction

Over the past decade, Selective Laser Melting (SLM) technology has grown at a rapid pace becoming the most used AM process for metal part fabrication. However, SLM poses challenges linked to the unexpected porosity defects generated in the components during the process, leading to a poor repeatability of the process, which have a direct influence on the structural integrity of the part (porosity, cracks, etc.) (Zhang et al.

T. Mabrouki et al. (Eds.): AIAM 2023, LNME, pp. 139–148, 2024.
https://doi.org/10.1007/978-3-031-47784-3_18

2017). Thus, a lack of consistent assessment of the quality of manufactured parts leads to costly post-process analysis, which can have an impact on the reproducibility of SLM technology. This is a major technological constraint, particularly in highly regulated sectors such as aerospace and the manufacture of medical devices.

To overcome this problem and improve SLM process monitoring, it is necessary to develop suitable control methods capable of providing rapid and reliable assessment of such defects. Numerous types of in-situ monitoring methods were investigated (Ydav et al. 2020, Peng et al. 2022), especially non-destructive testing (NDT) ones.

In the context of this study, the NDT EC system developed by AMiquam is employed to monitor and predict porosity defects in SLM-printed parts. Its ability to reliably sense global porosity rate has been demonstrated in recent studies (Spierings et al 2021, Spurek et al 2021) on 316L parts manufactured by SLM. This non-contact technique has the advantage of providing direct information on the mettalic surface while being less expensive, space-saving, and easy to install and calibrate than the other NDT methods usually used, such as X-ray tomography (CT) (Shihua et al 2022), acoustic signals (Pandian et al. 2021) or indirect measurements via optical monitoring systems of the molten bath using high-speed, high-resolution cameras (Bo et al 2021).

In this study, an in-situ automated quality control methodology of SLM is developed, by combining Machine Learning (ML) techniques with an empirical approach using non-destructive inspection measurements with eddy current sensors.

The approach gives the possibility of controlling the density of the part layer by layer. The suitability of combining ECT and ML for material porosity detection is demonstrated. The results underline the effectiveness of a such approach for SLM process.

2 Experimental Methods

2.1 SLM Dataset Collection: In-Situ Eddy Current Testing (ECT)

As shown in Fig. 1, an experimental set-up was established to ensure the in-situ monitoring of porosity defects generated in the parts during the process. The eddy current testing system (ECT system) developed by AMiquam (*AMiquam SA, Gland, Switzerland*) was integrated in the SLM Solutions 125 HL machine equipped with a single fiber laser (1x400 W). ECT measurements were carried out layer by layer on parts during the powder recoating process without modification of the build cycle.

To obtain multiple measurements during the printing process, two sensors (CH1 and CH2) were placed on the recoater (Fig. 1). These were initially positioned at 0.4 mm to the build plate surface.

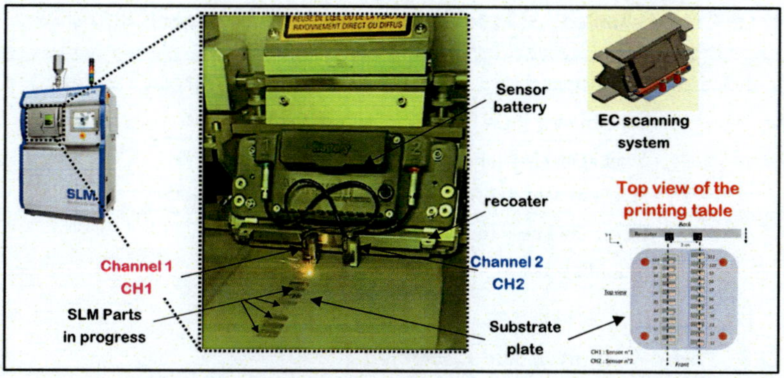

Fig. 1. Installation of EC system for in situ monitoring

2.2 Sample Manufacturing

Commercial Titanium alloy grade 23 ($TiAl_6V_4$) powder with a particle size ranging between 20 and 60 µm was used to additively manufacture a bulk sample with a size of 15 × 15 × 20 mm^3. Each part was subdivided into sub-pieces of 5 mm height which were printed with different parameters set composed of laser power (P); Scanning Speed (V); hatch distance (H) and scan strategy (Fig. 2).

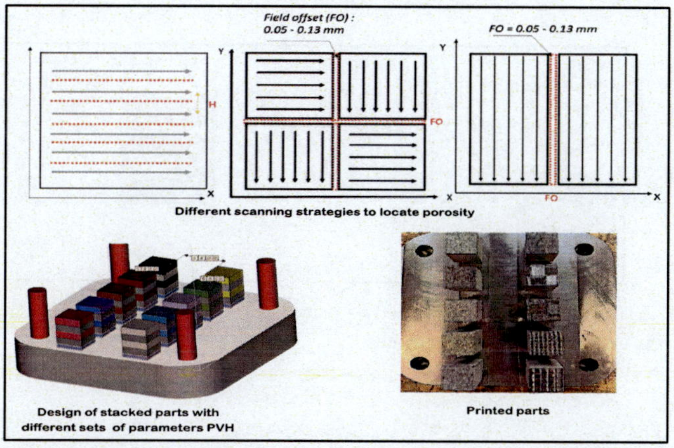

Fig. 2. Different geometrical configurations of pore insertion in SLM parts: a) by varying the scanning strategy (varying the FO), b) by level (sub-parts)

Process parameters ranges have been set so as to obtain parts with a variable porosity rate of up to 3%, and parts with localized porosity, through different scan strategies. Final parameters are presented in Table 1.

Table 1. SLM parameters applied to induce porosity in TiAl6V4

Parameters	Value interval
Laser power P (W)	100 - 300
Scan speed V (mm/s)	400 - 1800
Hatch distance H (mm)	0.075 - 0.25
Layer thickness (mm)	0.05
Shielding gas	Argon

2.3 Porosity Analysis and Labeling

After a preliminary testing step, different labels have been assigned to the sub pieces over time based on porosity rates which were quantified following two methods:

- Measurement of the relative global density using Archimedes method
- Measurement of localized porosity defects rate using metallographic analysis, followed by an image processing analysis based on a developed algorithm (Fig. 3). We were able to assess the porosity of each sub-piece (globally), and then we further refined the analysis to obtain the porosity of each layer.

Based on the methodology described above, a set of parameterized SLM tests has been conducted, thus constituting a database (DB) which will be used as the ground truth in Machine Learning algorithms detailed below.

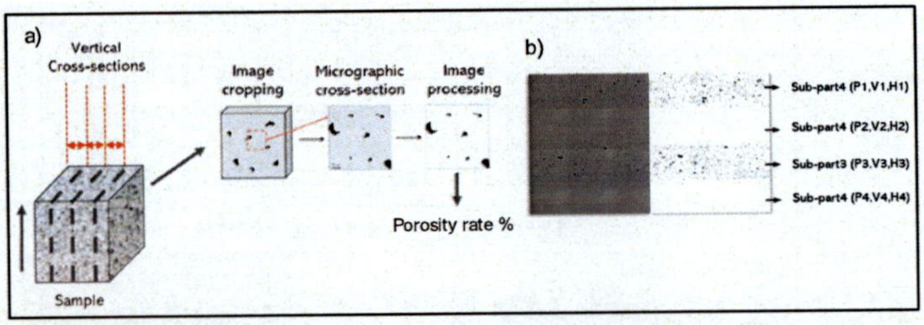

Fig. 3. a) Determination of localized porosity rates b) Example of metallography cliche of stacked sub-parts with different porosity rates.

2.4 Correlation

In order to correctly interpret the measurement signals, a correlation between porosity rate and the EC signal response was done. Figure 4 shows an example of the EC signal (imaginary and real signals) evolution with the porosity rate variation, during the manufacturing process, throughout a piece made up of 4 sub-parts printed with different SLM parameters. In the same context, (Spierings et al 2021) underlined, using the same EC instrument, a very high correlation coefficient ($r^2 = 0.997$) between the relative measured density over 316L cube samples obtained by SLM, and the EC sensor signal.

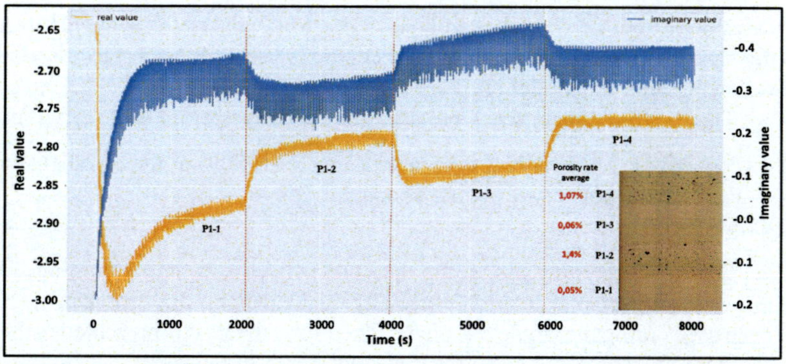

Fig. 4. Correlation between EC signal variation and measured porosity rate

3 Data Description

3.1 ECT Signal Processing Method

The collected ECT signals during the process were pre-processed to extract the useful values (real and imaginary components). Then, it was segmented by part and by layer using an implemented method based on the derivative of the signal itself. The raw sensor values were used directly Finally, the signal of each printed piece was post-processed to remove outliers, reduce noise and make it more stable. Figure 5 shows the final signals which will be used for machine learning.

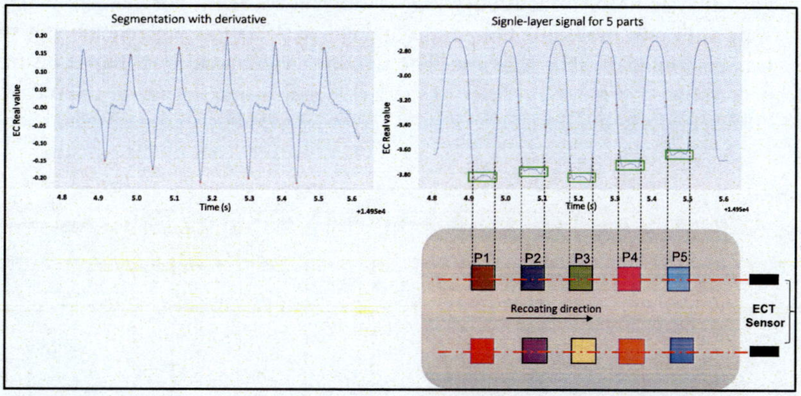

Fig. 5. Derived from the real signal for part separation and the real signal of each part.

3.2 Dataset Construction and Splitting

The whole data set contains 66 different pieces, most of them made up of 4 sub-parts, manufactured with different SLM parameters, giving a set of about 250 distinct test cases i.e., about 20′400 labelled layers. The whole dataset is available in *zenodo* under the identifier https://doi.org/10.5281/zenodo.7886192. It can be accessed directly at the following link: https://doi.org/10.5281/zenodo.7886192.

Three different data sets were used for the regression of these different porosities:

- Parts with a porosity less than or equal to 1% (~90% of the complete dataset)
- Parts with a porosity less than or equal to 3%
- All Parts

During the training phase, the dataset has been split into a training, validation and test sets using cross-validation especially the Leave-one-out approach (LOO) which has been used during the whole test phase. Therefore, the results presented in the next section show the average result of all the different usable samples.

4 Algorithms

Regression of the porosity rate of different samples was made with the EC signals and the different labels mentioned above in paragraph 2.4. According to porosity defect location to be estimated, three different approaches were explored: classical machine learning algorithms (SVM) and convolutional neural networks (CNN): CNN-LSTM and CNN2D.

4.1 Baseline: SVM with Statistical Description Windows

As a first attempt, a Support Vector Machine (SVM) was used for predicting the global porosity rate of manufactured Ti-6Al-4V pieces by SLM. For that, collected EC signals were split into multitude temporal windows of 25 layers where each statistical feature was computed as shown below. The window width was fixed based on the estimated penetration depth $\delta \approx 1.49$mm of the EC signal inside the metal (García-Martín et al. 2011) (Fig. 6).

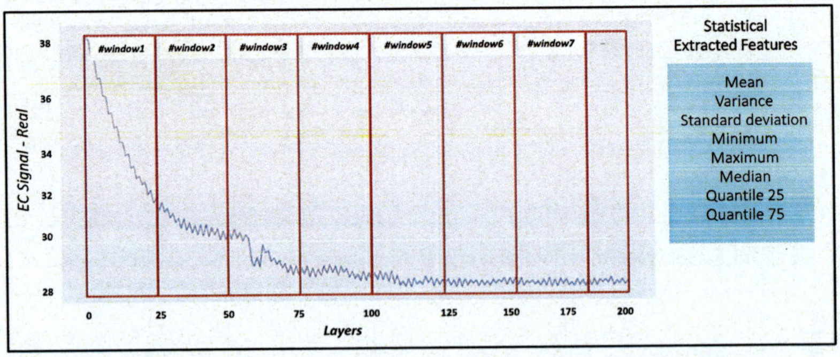

Fig. 6. Example of inputs for the SVM: Representation of fixed windows over the EC signal and statistical features extracted.

4.2 Neural Network: CNN – LSTM

In a second step, the CNN Long Short-Term Memory Network (CNN-LSTM) (Fig. 7-a) has been used for the regression of a more local porosity (each 100/50/25/12/1 layers of the piece) to give real-time feedback of the piece's porosity rate. Then the global porosity of the piece can be computed and compared to the SVM's results. Input of the neural net are no longer statistical features, but the signal itself. The estimations of the porosity of layer L needs the signals of the 29 next layers (*l, l + 1, ..., l + 29*, where *l* is the signal gathered at the layer).

The shape of the input data is: (number of samples, number of layers (30), number of values per layer (20), channel (2: real and imaginary)).

4.3 Neural Network: CNN2D

To improve the result of the previous neural network approach, a regular two-dimensional CNN (CNN2D), illustrated in Fig. 7-b, was applied. In this case, the temporality of the signal is handled by convolution and no more by the LSTM.

The input is the same as the CNN1D – LSTM neural network, except 15 layers before the targeted layer, are used in addition to the 30 next ones.

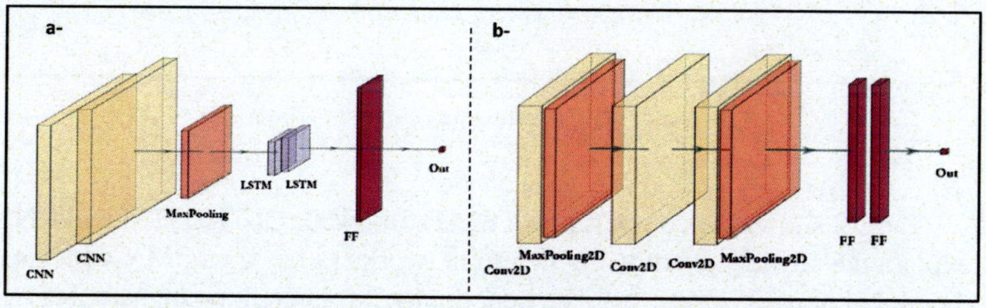

Fig. 7. Architecture of a- CNN-LSTM network, b- CNN2D network

5 Results and Evaluation

Figure 8 shows the results of the porosity regression for each sub-piece. As most of the values are below 0.5%, it tends to make good estimation for dense parts, which is not the case for higher porosity.

Given that the first layers are affected by the building plate, Fig. 9 shows an example of the porosity defect rate predicted layer-by-layer (Fig. 9-a) and through the global sample over time (Fig. 9-b) (Ex: sub-piece P1 of the part #N5 of the 20.01.21).

To quantify the regression error, the Mean Absolute Error (MAE) metric (Eq. 1) was calculated, it's defined as the average deviation between the predicted value $\hat{y_i}$ and the true y_i, , i.e., porosity rate [%].

$$MAE = \frac{1}{n}\sum\nolimits_{i=1}^{n} |\hat{y}_i - y_i| \quad (1)$$

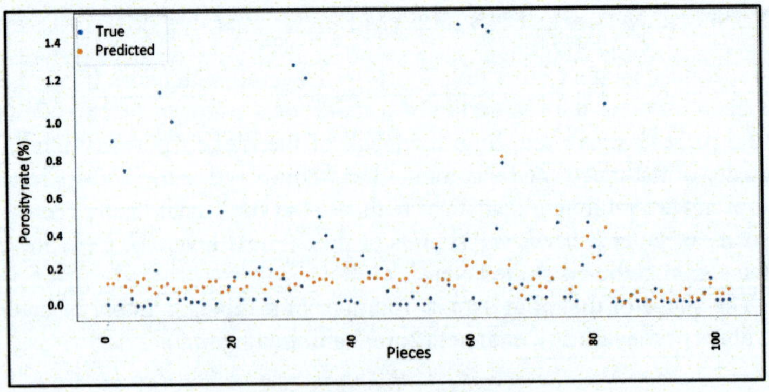

Fig. 8. Result of SVM estimations.

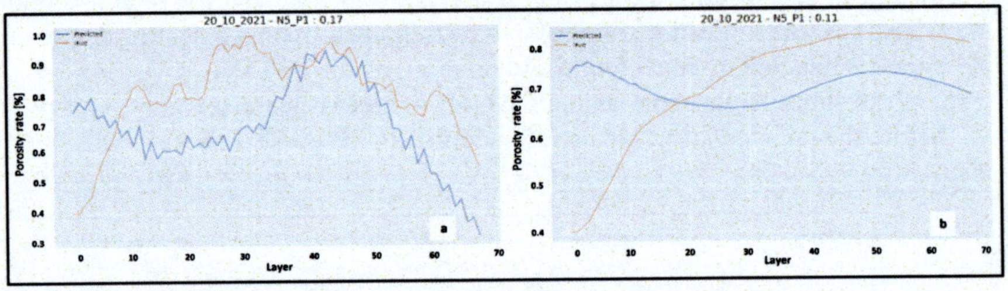

Fig. 9. Results of CNN estimations

Table 2 shows the comparison of MAE for SVM, CNN-LSTM and CNN2D. Results explain the advantages of using neural networks rather than SMV which estimates only the global porosity rate only. The benchmark has been done on samples with a porosity rate below 1%, which represents 90% of the available samples.

Table 2. AI porosity defects prediction

Models	*Test*		
	Layer-by-layer porosity [MAE]	Local porosity over time [MAE]	Global porosity [MAE]
SVM	–	–	0.21
CNN-LSTM	0.10	0.084	0.16
CNN2D	0.11	0.079	0.12

As the SMV estimate only the global porosity with statistical features, the only relevant comparison should only be based on this scenario. Results highlight a better accuracy of the CNN algorithms with a better prediction of the porosity of each layer.

CNN2D seems to be a little more accurate than CNN-LSTM. But overall, the two networks are quite similar.

Considering the current state of the art in situ monitoring using EC technique (Spurek et al 2021), the most significant outcoming of the present study is the algorithm ability to predict layer by layer porosity during the SLM process with an average prediction error of 0.1% porosity rate. This is an excellent result which clearly demonstrates the effectiveness of the proposed approach which can be used for parameters process optimization to automatically adjust them according to part quality.

6 Conclusion

This study investigates the feasibility of using eddy current testing (ECT) technique combining machine learning for automated in-situ quality monitoring of SLM process.

An experimental set-up was established by integrating the EC sensor on the SLM machine which has enabled to collect signals for the detection of porosity defects generated in Ti-6Al-4V parts during the fabrication. The correlation between the porosity rate and the average EC signal was established and it was decided to target porosity defects between 0 and 3%. A DB of labelled 204 cases, which correspond to 20400 layers, has been created. This data was used to input the AI developed algorithms, and then predicted the porosity rates layer by layer, as well as the overall porosity of the parts.

It was shown that porosity defects rate can be predicted layer by layer by the CNN2D and CNN-LSTM algorithms with an average prediction error MAE of 0.1% and 0.11% porosity rate, demonstrating the effectiveness of the proposed approach.

Acknowledgements. The authors like to thank HES-SO University of Applied Sciences and Arts Western Switzerland for financing the study within the frame of DEEPSLM project.

References

Zhang, B., Li, Y., Bai, Q.: Defect formation mechanisms in selective laser melting: a review. Chin. J. Mech. Eng. **30**, 515–527 (2017). https://doi.org/10.1007/s10033-017-0121-5

Yadav, P., Rigo, O., Arvieu, C., Le Guen, E., Lacoste, E.: In situ monitoring systems of the SLM process: on the need to develop machine learning models for data processing. Crystals **10**(6), 524 (2020). https://doi.org/10.3390/cryst10060524

Peng, X., Kong, L., An, H., Dong, G.: A review of in situ defect detection and monitoring technologies in selective laser melting. 3D Printing Addit. Manuf. **10**(3), 438–466 (2023). https://doi.org/10.1089/3dp.2021.0114

Spierings, A.B, Spurek, M.A, Lany, M, Santi, G., Revaz, B., Wegener K.: Direct part density inspection in laser powder bed fusion using eddy current testing. In: Joint Special Interest Group meeting between euspen and ASPE (2021). https://www.research-collection.ethz.ch/handle/20.500.11850/516133

Spurek, M.A., et al.: Relative density measurement of PBF-manufactured 316L and AlSi10Mg samples via eddy current testing. Metals **11**(9), 1376 (2021). https://doi.org/10.3390/met11091376

Shihua, W., Huiyuan, Q., Shengkai, Y., Su Xia, Z.: Nondestructive investigation on close and open porosity of additively manufactured parts using an X-ray computed tomography. Mater. Today Proc. **70**, 124–130 (2022)

Pandiyan, V., et al.: Semi-supervised monitoring of laser powder bed fusion process based on acoustic emissions. Virtual Phys. Prototyping **16**, 481–497 (2021). https://doi.org/10.1080/17452759.2021.1966166

Bo, E., Xiao-yuan, Z., Huan-qing, Y., Dong-jian, P., Ze-ming, W., Yuan-jie, W.: In situ monitoring methods for selective laser melting additive manufacturing process based on images–a review. China Foundry **18**, 265–285 (2021). https://doi.org/10.1007/s41230-021-1111-x

García-Martín, J., Gómez-Gil, J., Vázquez-Sánchez, E.: Non-destructive techniques based on eddy current testing. Sensors **11**, 2525–2565 (2011). https://doi.org/10.3390/s110302525

Spurek, M.A., et al.: In-situ monitoring of powder bed fusion of metals using eddy current testing. Addit. Manuf. **60**, 103259 (2022). https://doi.org/10.1016/j.addma.2022.103259

Non-conformities and Scrap Costs Reduction in a 3D Printing Workshop

Safa Mathlouthi[1,2](✉) and Abd El Hedi Gabsi[2]

[1] Applied Mechanics and Engineering Laboratory (LMAI), University of Tunis El Manar, National Engineering School of Tunis (ENIT-Tunisia), Campus El Manar, Tunis, Tunisia
safa.mathlouthi@enit.utm.tn

[2] Higher Institute of Technological Studies of Nabeul, Mrezga University Campus, 8000 Nabeul, Tunisia

Abstract. The aim of this study is to reduce non-conformities and scrap costs in a Tunisian 3D printing workshop. Therefore, we performed a resolution of problems methodology. We started by defining the problem. According to a preliminary diagnosis of the 3D printing workshop, it was impossible to conduct a numerical analysis due to the lack of indicators and non-conformities registration. So, we began our work by establishing key performance indicators to gain insights into the different machines of this workshop. Then, we developed a non-conformities registration document. This document was essential to realize the Pareto Analysis which revealed that" low extrusion density" is the main non-conformity. After identifying its main causes through a root cause analysis, we implemented improvement actions consisting principally of the optimization of printing parameters. In the last step "review of results", we noticed an average of 15% reduction in the non-conformity rate and scrap cost. We showed also, through this article, that low extrusion density non-conformity exists regardless of the generation of the machine and that key performance indicators are essential and can be adapted to reflect the real state of 3D printing workshops in a Tunisian industry.

Keywords: Additive Manufacturing · PLA · non-conformity · extrusion density · scrap

1 Introduction

Additive manufacturing refers to the process of creating three-dimensional objects or parts by building up successive layers of material. This technology is also known as 3D printing. Additive manufacturing allows for the creation of complex geometries that would be difficult or impossible to produce with traditional manufacturing methods (Gibson et al. 2021). It has numerous applications in various industries (Baletti et al. 2017). Fused Deposition Modelling FMD is one of the additive manufacturing technologies used in 3D printing. Various polymers are used in FMD. Polylactic acid, or PLA, is the primary raw material commonly used in FMD 3D printing. It's a biodegradable

T. Mabrouki et al. (Eds.): AIAM 2023, LNME, pp. 149–158, 2024.
https://doi.org/10.1007/978-3-031-47784-3_19

thermoplastic polymer that is made from renewable resources such as corn starch, tapioca root, and sugarcane, and it's characterized by its low toxicity, ease of use, relatively low melting temperature, low cost, and availability (Turner et al. 2014). "Low Extrusion density" is one of the major non-conformities in 3D printing workshops treated in the literature. It refers to a situation where the printer is not extruding enough material during the printing process. (Fernando and Hélio 2023) studied the impact of the printing nozzle temperature variation on the mechanical and dimensional characteristics of a low-density PLA like low extrusion density problem. (Anis and Kamil 2021) discusses how variations in print speed and extrusion temperature affect the properties of PLA prints. These properties could include dimensional accuracy, mechanical strength, or surface quality (Soares et al. 2018) such as low extrusion density.

The main objective of this paper is to reduce scrap costs and non-conformities in a 3D printing workshop in a Tunisian case through the reduction of "low extrusion density" non-conformity. The purpose of this article is also to answer the following questions: Is "low extrusion density" non-conformity presents a major non-conformity in Tunisian 3D printing workshops? Is this problem appearance varying from one generation to another of 3D printing machines? Are Key Performance Indicators sufficient for Measuring 3D printing problems?

2 Characteristics of 3D Printing Machines

This work was elaborated in an industrial company specializing in the automotive industry. The company has a 3D printing workshop consisting of 11 machines of two different generations (8 machines CreatBot DE Plus and 3 machines PRUSA MKI3). The technical characteristics of these machines are summarized in Table 1.

Table 1. Technical characteristics of machines CreatBot De plus and PRUSA MKI3

Machines	Characteristics				
	Maximum print speed	Nozzle Diameter (default)	Filament Diameter	Software	Generation
CreatBot DE Plus	120 mm/s	0.4 mm	1.75 mm; 2.85 mm; 3.0 mm	CreatWare multilanguage	2011
PRUSA MKI3	200 mm/s	0.4 mm	1.75 mm	SlicerPRUSA Edition	2019

3 Methodology

To treat the problem, we followed this methodology (Fig. 1):

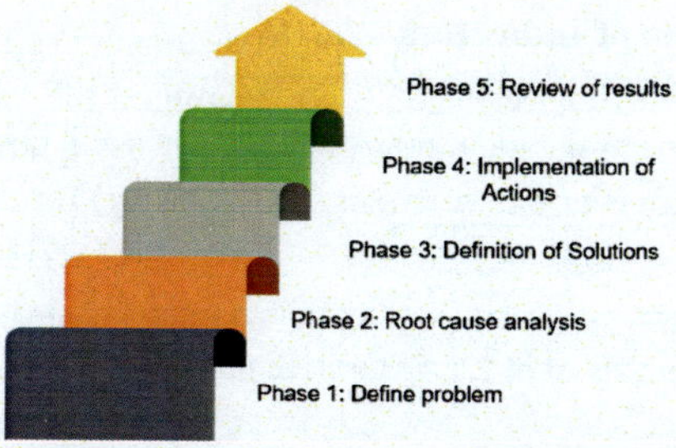

Fig. 1. Methodology of resolution of problems adopted

3.1 Define the Problem

The 3D printing workshop suffers from recurring shutdowns and various problems of quality (non-conformities and scrap). To precisely define the problem, we began our work by reporting the initial state of the workshop making floor observations (Gemba), and by analyzing the information system.

By visual diagnosis, we noticed that there is an important number of defective parts in the non-conformity zone having a "low extrusion density" issue and a lack of "non-conformities" registration document. Through an analysis of the information system, we discovered a lack of Key Performance Indicators. Only scrap quantities were registered.

So as first improvements, we started by implementing the following actions:

- Definition of Key Performance Indicators KPI
- Calculation and analysis of the defined indicators
- Implementation of the non-conformities registration document
- Analysis of non-conformities registration document

3.1.1 Definition of Smart Indicators

We started our approach by defining the following indicators.

- The synthetic efficiency rate TRS, also called OEE Overall Equipment Effectiveness, corresponds to the time during which the equipment is supposed to produce without considering planned shutdowns but integrating all the unpredictable and inherent events of automated production: failures, micro-stops, loss of cadence, non-quality, etc. This rate reflects the ratio of time used to produce quality out of planned stops.

$$TRS = \frac{Actual\ production\ time}{planned\ production\ time} \tag{1}$$

- Non-conformity Rate: This indicator points out the percentage of non-conforming parts which are principally discarded or thrown.

$$non\ conformity\ rate = \frac{Number\ of\ non\ conforming\ parts}{Total\ number\ of\ pieces\ produced} \tag{2}$$

- Scrap cost: it's about calculating the cost of thrown pieces.

$$Scrap\ cost = weight\ of\ thrown\ pieces\ on\ grams\ x\ price\ of\ one\ gram \tag{3}$$

3.1.2 Calculation of Indicators

During the month of October 2022, we examined the various performance indicators to pinpoint the machines encountering most of the quality issues and shutdowns. We calculated indicators for one filament spool (2.2 kg) for each machine (Table 2).

Table 2. Key Performance Indicators in the additive manufacturing workshop during October 2022 for one filament spool of 2.2 kg for each machine

Indicators	Machines					
	CreatBot DE Plus			PRUSA MKI3		
	PDE 02	PDE 03	PDE 04	PPA 01	PPA 02	PPA 03
TRS (%)	69	58	34	84	82	85
Non-conformities rate (%)	10	15.87	23.63	5	8.42	8.69
Scrap cost (euro)	22	23.76	48.62	9.38	3.94	7.07

3.1.3 Analysis of Indicators

By comparing the performance indicators, the CreatBot DE Plus machines (old generation) are less productive than the PRUSA MKI3 machines (new generation). In fact, the synthetic efficiency rate of the CreatBot DE Plus machines varies from 34% to 69%, whereas the PRUSA MKI3 machine's rate ranges from 82% to 85%. The PDE 04 machine is considered to have the lowest efficiency and the highest non-conformities rate and scrap cost among the CreatBot DE Plus machines. Within the PRUSA MKI3 machines, the PP01 machine has the lowest non-conformities rate but the highest scrap cost since the weight of the rejected parts is the biggest.

3.1.4 Implementation of Non-conformities Document

Due to the lack of traceability about the types of non-conformities across all 3D printing machines, we began by implementing a new document, in which the types of non-conformities, the number, and the weight must be notified.

The non-conformities types indicated in this document are as follows: Drooling, delamination on the side or the top, not enough material on the thin parts, one layer moving horizontally, thin walls delaminating, the angles bending upwards, the corners coming loose, low extrusion density, layers are poorly welded, bubbles form on the first layer, the tops of the holes are collapsing, the diapers are delaminating, droplets appear, an extruder Bowden drooling.

This document will present our database in the analysis of non-conformity types.

3.1.5 Analysis of Non-conformities Registration Document

Through a Pareto analysis of the registration document already in use for a month (November 2022), we concluded that "low extrusion density" is the primary issue with 92% of non-conformities. This analysis (regardless of the machine generation) is presented in the Fig. 2.

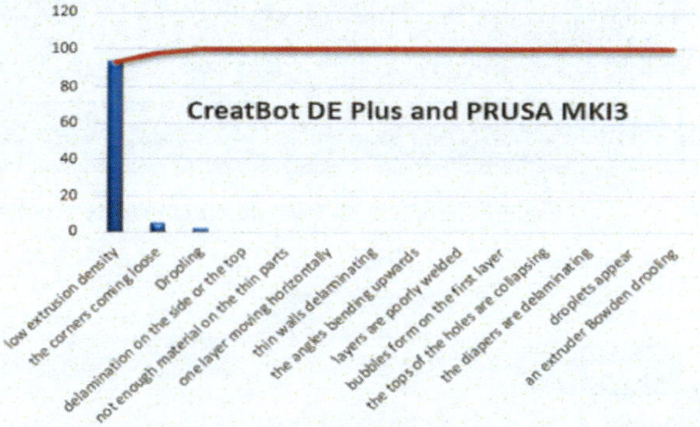

Fig. 2. Pareto Analysis of Non-conformities for the Month of November 2022

To better analyze the appearance of this non-conformity per generation of machine, we made a Pareto analysis of non-conformities for CreatBot DE Plus and PRUSA MKI3 separately (Fig. 3).

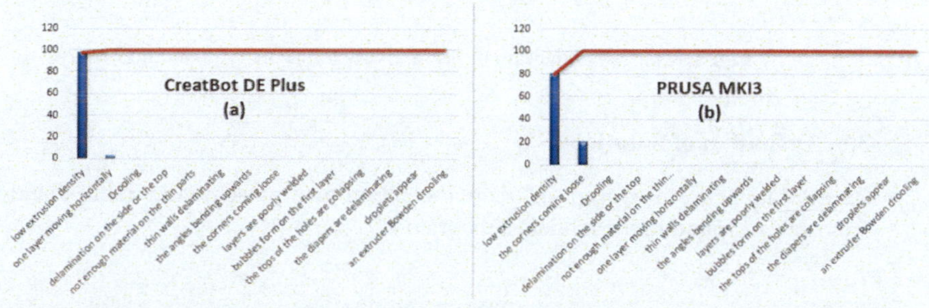

Fig. 3. Pareto Analysis of Non-conformities for the Month of November 2022: (a) CreatBot DE Plus (b) PRUSA MKI3

Figure 4 shows a sample with a "low extrusion density" issue.

"Low extrusion density" can result in incomplete or weak print gaps between layers, and other defects in the final product.

3.2 Root Cause Analysis

This step consists of searching the different causes of low extrusion density non-conformity.In our case, we used the Ishikawa method as shown in Fig. 5.

Through this diagram, we conclude that " Low extrusion density "can be caused by several factors such as filament diameter and quality, machine model, extruder tension, retraction setting, print speed, nozzle temperature …

Fig. 4. Sample with low extrusion density problem

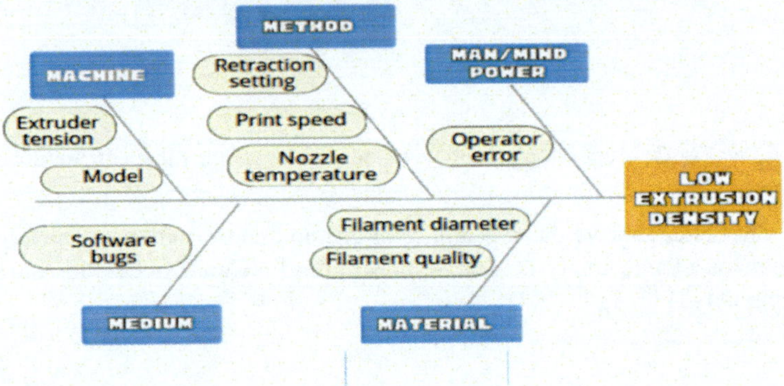

Fig. 5. Ishikawa Diagram of "Low Extrusion Density" non-conformity

3.3 Definition of Solutions

As a solution, we proposed to follow the action plan presented in Table 3 and issued from causes found in the Ishikawa diagram.

Table 3. Plan Actions

Causes	Action	Date /period
Nozzle temperature	Optimization of nozzle temperature	December 2022
Print speed	Optimization of print speed	January 2023
Retraction speed	Adjustment of the retraction setting	January 2023
Filament diameter and quality	Optimization of filament diameter	January 2022
	Optimization of filament quality	January 2022
Extruder tension	Adjustment of extruder tension	January 2023
Software bugs	Use of other software	January 2023

To guarantee better results, we also proposed other actions relative to maintenance and calibration as shown in Table 4.

Table 4. Maintenance Actions

Action	Date /period
Calibration of Extruder	December 2022
Change of the extruder position	January 2023
Calibration of the printing bed	December 2022
E-steps calibration	December 2022
Change of the software	December 2022

3.4 Implementation of Improvement Actions

3.4.1 Optimization of Nozzle Temperature

Depending on the brand and the type of filament and according to manufacturer documentation, the optimal nozzle temperature for PLA can vary between 180 to 230 °C. In our case, we tried to vary the nozzle temperature in the recommended interval and track low extrusion density non-conformity. This action was done on each generation of machines (CreatBot DE Plus and PRUSA MKI3) for a filament spool of 2.2 kg. The best nozzle temperature concluded is 210 °C for CreatBot DE Plus and 190 °C for PRUSA MKI3.

3.4.2 Optimization of Print Speed

The ideal print speed for PLA can vary between 20–60 mm/s, depending on the layer thickness and other factors. In our case, we tried to vary the print speed in the recommended interval and track low extrusion density non-conformity. This action was done on each generation of machines (CreatBot DE Plus and PRUSA MKI3) for a filament spool of 2.2 kg. The best print speed for both machines is 60 mm/s.

3.4.3 Adjustment of the Retraction Setting

This action consists of adjusting the retraction settings to prevent clogging and improve the overall quality of the print. The optimal retraction distance for PLA can vary between 1–2 mm, while the retraction speed can vary between 30–60 mm/s. For each generation of machines (CreatBot DE Plus and PRUSA MKI3), we varied both retraction distance and retraction speed.

We deduced that for both machines, the best retraction setting is 1 mm for the retraction distance and 30 mm/s for the retraction speed.

3.4.4 Optimization of Filament Diameter and Quality

We tried first to check that the filament diameter fits the printer's settings: 1,75 mm PRUSA MKI3 and 2,85 mm CreatBot De Plus. We tried then to verify that the filament is of high quality. In fact, we used one spool of 2.2 kg on each machine from three different suppliers and compared the obtained quality for the final products. We noticed that for the first supplier (the actual one), 23% of produced parts have low extrusion density non-conformity, for the second supplier 17%, and for the third supplier only 2% of final products have this issue. So, we concluded that changing the supplier of filament from the actual supplier (N°1) to the supplier N°3 can reduce the low extrusion density non-conformity.

3.4.5 Adjustment of the Extruder Tension

This action consists of adjusting the extruder tension to ensure that the filament is being properly fed into the printer. The optimal tension for PLA can vary depending on the specific printer and filament being used, but it's important to ensure that the tension is neither too loose nor too tight.

3.4.6 Maintenance and Calibration Actions

Calibration actions are critical for achieving accurate and consistent prints. In fact, it's essential to periodically recalibrate the extruder and other components to account for wear, changes in environmental conditions, or upgrades to the printers.

It aims to ensure that our machines' components are properly adjusted and functioning as intended.

The extruder calibration focuses on obtaining the correct filament flow rate and ensuring proper filament extrusion.

E-steps, or the number of steps the extruder motor takes to extrude a certain length of filament, require calibration. This calibration ensures that the extruder accurately moves the required quantity of filament.

Other than calibration actions, we made some changes in components as regular maintenance to contribute to consistent print quality.

3.5 Review of Results and Discussion

Due to the improvement actions, the performance indicators in the 3D printing machines have been improved.

In fact, we present in this table the values of indicators one month after the implementation of the actions.

The indicators below are for one filament spool of 2.2 kg for each machine.

Table 5 demonstrates that practically TRS indicators have been improved by 15% for all machines.

Comparing TRS indicators, it's obvious that the PRUSA MKI3 machines still have better performance than that of CreatBot DE Plus machines.

The non-conformities rate has been reduced also by 15% (from 14.18% to 15.54%), making it clear that we met our primary goal.

Table 5. Key performance indicators of machines after improvement actions implementation during March 2023 for one filament spool of 2.2 kg

Indicators	CreatBot DE Plus			PRUSA MKI3		
	PDE 02	PDE 03	PDE 04	PPA 01	PPA 02	PPA 03
TRS (%)	79	67	39	96	94	97
Non-conformities rate (%)	8,57	14.28	20	4.16	7.36	7.82
Scrap cost (euro)	18.86	21.39	41.147	7.82	3.45	6.371

Due to the amelioration in our performance indicators, the scrap cost has been reduced for all machines (reduction from 9 to 16.6%); for example, by 16.6% for PPA01 and by 15.37% for PDE04.

In our case, we implemented all the actions at once for two months (due to a scheduling issue), but we also scheduled an artificial intelligence study to assess the impact of each parameter on the decrease of "low extrusion density" non-conformity.

4 Conclusion

In the present study, we searched to reduce non-conformities and scrap costs in a Tunisian 3D printing workshop. By conducting a thorough analysis of the root cause of the main non-conformity "low extrusion density", we were able to successfully implement improvement measures and achieve significant results. Our findings showed an average of 15% reduction in non-conformity rates and a reduction in associated costs, with a percentage decrease from 9% to 16.6%.

We concluded also that low extrusion density is a major issue in Tunisian industries which exists regardless of the generation of machines.

To further improve the results, we recommend conducting an artificial intelligence study to evaluate the influence of changing printing parameters on reducing the low extrusion density non-conformity for both machines (CreatBot DE Plus and PRUSA MKI3). Additionally, we may be able to compare the effects of these parameters between the two generations of machines.

Further research can build on our findings to continue improving the quality and efficiency of the 3D printing process.

Acknowledgment. We would like to thank SAME Tunisia for providing us with the resources necessary to conduct this work. We are grateful for their support and for the opportunity to collaborate with them on this project.

References

Balletti, C., Ballarin, M., Guerra, F.: 3D printing: state of the art and future perspectives. J. Cult. Herit. **26**, 172–182 (2017)

Turner, N.B., Strong, R., Gold, S.A.: A review of melt extrusion additive manufacturing processes: i. Process design and modeling. Rapid Prototyping J. **20**(3), 192–204 (2014)

de Freitas, F., Pegado, H.: Impact of nozzle temperature on dimensional and mechanical characteristics of low-density PLA. Int. J. Adv. Manuf. Technol. **126**(3–4), 1629–1638 (2023). https://doi.org/10.1007/s00170-023-11236-0

Ansari, A.A., Kamil, M.: Effect of print speed and extrusion temperature on properties of 3D printed PLA using fused deposition modeling process. Mater. Today Proc. **45**, 5462–5468 (2021). https://doi.org/10.1016/j.matpr.2021.02.137

Soares, J.B., Finamor, J., Silva, F.P., Roldo, L., Cândido, L.H.: Analysis of the influence of polylactic acid (PLA) color on FDM 3D printing temperature and part finishing. Rapid Prototyping J. **24**, 1305–1316 (2018)

Gibson, I., Rosen, D., Stucker, B., Khorasani, M.: Additive manufacturing technologies. Springer, Boston, MA (2010). https://doi.org/10.1007/978-1-4419-1120-9

Contribution to the Formulation of a Material Based on Marble Waste for Its Use in 3D Printing in Civil Engineering

Habib Zargayouna[1(✉)], Essaieb Hamdi[1], and Tarek Mabrouki[2]

[1] LR-ING-ENIT, University of Tunis El Manar/National School of Engineers of Tunis, BP 37 Le Belvédère, 1002 Tunis, Tunisia
zargayounahabib@gmail.com

[2] LR-MAI-ENIT, University of Tunis El Manar/National School of Engineers of Tunis, BP 37 Le Belvédère, 1002 Tunis, Tunisia

Abstract. The objective of this research is to experimentally study the way of recovery and recycling of marble processing waste mixed with other materials for 3D printing in Civil Engineering. In this paper, we present the framework of this study and the approach to be followed in the feasibility study of the use of Tunisian marble powder in the formulation of a printable material in Civil Engineering. Local materials from Tunisia (marble rock processing waste), now considered as waste, will be recycled in the field of additive manufacturing. We propose an experimental protocol and the first results of the characterization of the raw materials used. In particular, the marble powder used is calcium-dominant and contains a low percentage of Quartz and Dolimite, It has a density of 1.22.

Keywords: 3D printing · Marble · Additives · Civil construction · Waste

1 Introduction

3D printing in Civil Engineering presents an interesting alternative to respond to the increase in population and urbanization. It is indeed an emerging technology that requires more scientific research efforts for its implementation. In this context, the recovery of processing-induced materials resulting from the transformation of natural rock, whether ornamental or not, encourages researchers to find solutions to solve the problem of the management of this waste in view to automating the construction.

3D printing, also called additive manufacturing, can be used to produce architectural elements such as walls, columns, houses and entire buildings. This technology offers several advantages for the construction industry, including speed of production, precision, design and cost reduction.

Several studies have focused on the formulation of different types of construction materials. For example, extrusion printing is performed by depositing layers of fresh concrete. It depends on concrete properties such as pumpability, extrudability, buildability [1]. Extrudability is the ease of extruding concrete through the nozzle at a given

T. Mabrouki et al. (Eds.): AIAM 2023, LNME, pp. 159–169, 2024.
https://doi.org/10.1007/978-3-031-47784-3_20

speed. Constructability is the ability of stamped concrete to retain the imposed loads of subsequent layers without deformation [2].

Yield strength is the most important material parameter that governs both extrudability and constructability [3]. If the material in the lower layers has a very low yield strength, it may not have enough strength to withstand stresses from its own weight as well as the weight of the layers deposited above. On the other hand, a very high yield strength makes the extrudability of the material more difficult. Therefore, there is an optimum range for the yield strength in the case of pumping and extrusion of concrete [4]. Wangler et al. (2019) [5] have sorted out a few important research areas considering the components and parameters of the mixture. Rheological parameters such as yield strength, plastic viscosity and thixotropy also emphasized a relationship between processing and material performance, the effect of admixtures and their interaction on rheological and mechanical properties. Table 1 presents several works that have been interested in the design of 3D printable mixtures for different types of materials. Nair et al. (2019), [6] linked the rheological properties to the design of the mixing and extrusion geometry to the properties of the microstructure of the fresh mix. Compositional aspects such as particle size distribution, presence of superplasticizer, etc. were also considered by Nair et al. (2019) [6]. Wolfs et al. (2019a) [7] experimentally investigated the relationship between process parameters such as interlayer time interval, nozzle height and surface dehydration, and interlayer bond strength. Mechanical properties such as compressive strength and flexural strength were also studied with the three perpendicular directions. The influence of the drying of the printed layer and the time interval between layers on the decrease in bond strength has been discussed. Le et al. (2012b) [1] examined the hardened properties and fresh properties of high performance concrete with strength tested parallel and perpendicular to layer orientation, and bond strength for different time intervals between impressions.

The objective of this research is to experimentally study the way of recovery and recycling of Tunisian marble processing waste mixed with other materials for 3D printing. It is both an effective means in the face of the economic and ecological constraints of recent years and also an effective means of managing marble waste which constitutes a real and unavoidable problem.

2 Methodology

This research focuses on the characterization of mixtures of marble waste with different percentages of gypsum in order to prepare an adequate formulation for 3D printing. The chosen formula will then be tested in order to determine its resistance to the various mechanical stresses as well as its durability when subjected to weather conditions and thermal or acoustic insulation. The use of sustainable and ecological building materials is an important issue for the construction industry. With this in mind, the use of marble waste in the manufacture of building materials is increasingly being explored. The methodology of this study will include several steps. First of all, it will be necessary to collect samples of marble waste from different production sites. The samples will then be prepared and characterized using advanced characterization techniques such as X-ray diffraction, infrared spectroscopy and scanning electron microscopy. The physical and

Table 1. 3D printable blend design

3D printable mix design	References
A cement of lunar anorthite and glass is made by sintering to build dwellings on the moon	[8]
3:2 sand/binder, 0.26 water/cement and 20% fly ash and 10% silica fume and polypropylene fibers	[1]
CEM I 52.5 R + a limestone filler Aggregates (siliceous) of size less than 1 mm and polypropylene fibers are used	[7–10]
Developed "sleeping" mortar, which is delayed at the time of mixing and accelerated at the print head for rapid setting	[5]
Foam concrete for 3D printing is developed with a density range of 970 to 1500 kg/m3	[11]
The cementitious binder of pre-printing and post-printing parameters such as steady state extrudability pressure, stability, mechanical properties and dimensional accuracy of the print	[12]
A Yield Strength Based Approach for 3D Printed Concrete; Nano clay and silica fume	[4, 13]
Alternative binders (geopolymer concrete), and nano-clay as an additive	[14]
Durability study using magnesium potassium phosphate cement with 0–60% fly ash replacement and 10% silica fume replacement is performed	[15]
Martian concrete is made from a combination of sulfur and a replica of martial sol	[16]
Portland cement and class F fly ash + quartz sand less than 2 mm + quartz powder less than 100 microns + water-cement ratio of 0.32 + superplasticizer Printer bed size is 40 × 30 cm can print 30 cm height. The printhead has a rectangular nozzle 30 × 20 mm	[4, 13]

mechanical properties of the samples will also be measured to assess their suitability for 3D printing. Next, marble waste material formulations will be developed using different combinations of marble waste and binders. The formulations will be tested to determine their viscosity, stability, setting time…etc. The most promising formulations will be selected for 3D printing tests. 3D printing tests will be carried out using a 3D printer suitable for construction. Materials will be printed using different print setups to assess their feasibility and performance.

2.1 Valorization Alternatives of Marble Powder for 3D Printing

Some of the marble powder valorization alternatives in 3D printing include:

- Manufacture of decorative pieces
- Printing architectural parts
- Furniture printing
- Production of wall coverings

- Production of parts for civil engineering

In any case, the use of marble powder in 3D printing makes it possible to recover marble waste while offering infinite design and customization possibilities. In addition, by using recycled materials, 3D printing helps reduce the environmental footprint of industrial production.

2.2 Gypsum Properties

There are several types of gypsum and their properties can vary depending on their geological origin and method of extraction. With regard to the Tatouine gypsum, Tunisia, here are some general properties associated with them:

- Purity: Tatouine gypsum is considered to be of high purity, with a $CaSO4$ (calcium sulphate) content of over 95%.
- Color: They have a white to light beige color.
- Texture: their texture is fine and powdery, with a grain size ranging from 100 to 800 microns.
- Resistance: Tatouine gypsums have low compressive strength, which makes them easier to work with for applications such as construction or 3D printing.
- Hydration: When exposed to water, Tatouine gypsum has the ability to hydrate and harden into plaster.
- Reactivity: Tatouine gypsums are reactive with other materials such as lime, cement and marble powder, which makes it possible to create mixtures for specific applications.

2.3 Limes Properties

Limes are widely used materials in the construction industry in Tunisia. Here are some of their common properties:

- Purity: the lime produced in Tunisia is generally considered to be of high quality, with a high content of calcium oxide (CaO) above 90%.
- Colour: powdered lime has a white to beige colour.
- Reactivity: lime is highly reactive with water, which allows it to be transformed into a paste used for mortars and plasters.
- Carbonation: lime has the property of carbonating, i.e. reacting slowly with carbon dioxide in the air to form calcium carbonate, which makes mortars and plasters more resistant.
- Hardness: limes are relatively soft and hard at the same time, which allows them to bond with other materials and mold easily.
- Strength: Limes are not as strong as some other building materials, such as cement, but they are strong enough to be used in applications such as mortars and plasters.
- Varieties: There are different types of lime, including air lime (non-hydraulic), natural hydraulic lime and artificial hydraulic lime, each with different properties and uses.

2.4 Chemical Reaction Between Marble Powder and Gypsum

When marble powder and gypsum are mixed together in the presence of water, a chemical reaction called setting reaction occurs, which forms a hard, solid material called plaster. The chemical reaction between marble powder and gypsum can be described as follows:

$$CaSO4.2H2O(gypsum) + CaCO3(marble\ powder) + 2H2O \rightarrow 2CaSO4.2H2O(plaster)$$

In this reaction, gypsum (CaSO4.2H2O) reacts with marble powder (CaCO3) to form calcium sulphate dihydrate (CaSO4.2H2O), which is the main component of plaster. Water is needed to initiate the reaction and to allow the formation of hardened plaster. The plaster thus formed is used in a variety of applications, including making moldings, repairing drywall, covering walls and ceilings, etc.

2.5 Chemical Reaction Between Marble Powder and Lime

The chemical reaction between marble powder and lime (or calcium oxide) is an exothermic reaction that produces calcium carbonate and water. This reaction is called the carbonation reaction or the hydraulic lime formation reaction. The chemical reaction between marble powder and lime can be described as follows:

$$CaO\ (lime) + CaCO3\ (marble\ powder) + H2O \rightarrow 2CaCO3\ (calcium\ carbonate) + heat$$

In this reaction, the lime reacts with the marble powder to form calcium carbonate, which is a solid, crystalline material. Water is also produced during this reaction. This reaction is exothermic, which means that it releases heat. The calcium carbonate thus formed can be used in a variety of applications, including the production of cements, the manufacture of glass and ceramics, the production of paper pulp, etc. The carbonation reaction is also important in the conservation of limestone buildings, as it forms a protective layer of calcium carbonate on the surface of the stone, which can help protect the stone from air pollution and weathering degradation.

3 Material Characterization Tests for 3D Printing

3.1 Marble Powder Characterization Tests for 3D Printing

Marble powder characterization tests may include the following tests:

- Particle size analysis: This analysis makes it possible to determine the particle size distribution of the marble powder. It is carried out using a mechanical sieve or a laser method to measure the particle size.
- Chemical analysis: This analysis determines the chemical composition of the marble powder, including the levels of calcium, magnesium and other elements.
- Morphological analysis: This analysis examines the shape of the marble powder particles, which can help understand their behavior when used in 3D printing.
- Density analysis: This analysis helps to determine the density of the marble powder, which can have an impact on the properties of the 3D printing.

- Porosity analysis: This analysis helps to determine the level of porosity of the marble powder, which can also affect the properties of the 3D printing.
- Strength analysis: This analysis measures the mechanical strength of marble powder, which is important for understand its properties as a 3D printing material.

3.2 Gypsum Characterization Tests for 3D Printing

Characterization of gypsum for 3D printing may include the following tests:

- Particle size analysis: determine the size of the gypsum particles in the powder. It can be carried out using sieving or laser granulometry.
- Chemical analysis: determine the chemical composition of the gypsum, in particular the percentages of calcium, sulfur and oxygen. It can be performed using an X-ray fluorescence spectrometer.
- Measurement of apparent density: determine the bulk density of gypsum, which can vary depending on the particle size and shape of the particles. It can be done by using a graduated cylinder and measuring the mass of a given volume of gypsum.
- Setting time test: time required for the gypsum to solidify in the presence of water. Knowing this setting time is important for handling and using the material.
- Thermogravimetric analysis: measure the variation in mass of a gypsum sample as a function of temperature. It can be used to determine the water content of gypsum, as well as other thermal properties.
- Measurement of compressive strength: determine the compressive strength of gypsum.

3.3 Lime Characterization Tests for 3D Printing

There are several types of lime, each with different properties. However, here are some characterization tests that may be relevant for the use of lime in 3D printing in civil engineering:

- Chemical analysis: This analysis can give information on the chemical composition of the lime, such as the content of calcium oxide (CaO), magnesium oxide (MgO), silicon dioxide (SiO2) and alumina (Al2O3).
- Specific surface: The specific surface of lime can give information on the surface available for chemical reactions and interactions with other materials. The measurement of the specific surface can be carried out using different techniques, such as the BET method (Brunauer-Emmett-Teller) or the Blaine method.
- Particle size: Uniform particle size can help ensure even material distribution and reduce print nozzle clogging issues.
- Compressive strength: Compressive strength is an indicator of lime strength. It can be determined by carrying out compression tests on lime samples.

4 Materials and Methods

The methodology of this study will include several steps. First, it will be necessary to collect samples of marble waste from different production sites (Fig. 1). The samples will then be prepared and characterized using advanced characterization techniques such as X-ray diffraction (Fig. 2), infrared spectroscopy and scanning electron microscopy. The physico-chemical and mechanical properties of the samples will also be measured to qualify their suitability for 3D printing.

Figure 2 shows the XRD spectrum obtained for the marble powder used from the COGEMAC factory. It follows that this powder is mainly formed of calcite which is consistent with the genesis of the Tunisian marble rock which is the result of contact metamorphism of limestone.

3D printing tests will then be carried out using a 3D printer suitable for construction. Materials will be printed using different print setups to assess their feasibility and performance.

4.1 Robot Manufacturing Process for 3D Printing in Civil Engineering

Manufacturing a robot for 3D printing in civil engineering can be a complex process that involves several steps, including:

- Design: This is the phase where the specifications of the robot are defined according to the needs of the application. This includes size, shape, load capacity, speed and precision of movement.
- Assembly: In this phase, the different parts of the robot are assembled. This includes installing the base structure, motors, sensors, arms and print nozzles.
- Programming: The robot must be programmed to perform the tasks specific requirements for 3D printing. This involves configuring print parameters, such as print speed, nozzle temperature, material pressure, and layer height.
- Tests and adjustments: Once the robot is assembled and programmed, it must be tested. Adjustments may be needed to improve print quality, accuracy, or speed.

4.2 Experimental Protocol

The experimental additive manufacturing protocol proposed for 3D printing in civil engineering consists of:

- The collection and preparation of marble waste samples: Marble waste must be collected from factories and marble processors. In this research we have started sampling with industrial partners such as Marble factories: BENNOUR, KALLEL and COGEMAC.
- The preparation of the marble powder: which must be sieved to remove impurities and obtain a uniform particle size. Next, the marble powder must be dried to remove all moisture.
- Formulation of the bonding matrix: The bonding matrix must be formulated according to the mechanical properties desired for the final product.
- Mixing the marble powder and the bonding matrix: The marble powder must be thoroughly mixed with the bonding matrix to ensure an even distribution of the powder in the bonding matrix. The mixing ratio must be carefully chosen to obtain the desired mechanical properties.
- Preparation of the printing nozzle: The printing nozzle should be chosen according to the particle size of the marble powder and the viscosity of the bonding matrix. The nozzle should be cleaned and prepared before printing.

- 3D printing: Printing must be done with a digital control to ensure control of the printing speed. Nozzle temperature should be controlled to prevent clumping or other flow problems.
- Post-processing: After printing, the product should be post-processed to remove any remaining moisture and strengthen the mechanical properties of the material.

Fig. 1. Sampling procedure at COGEMAC

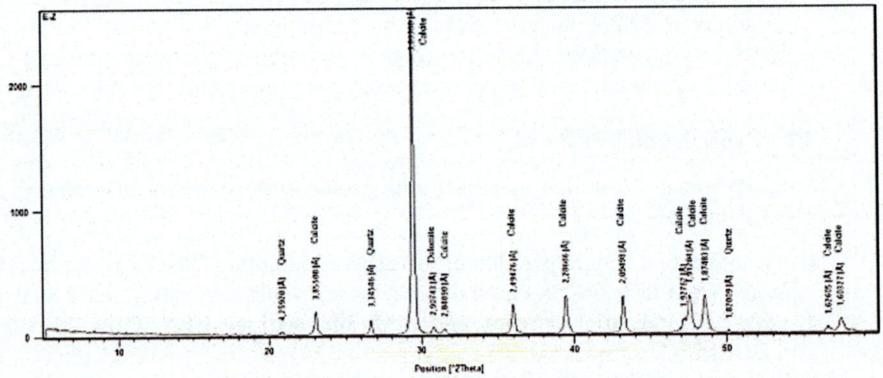

Fig. 2. XRD of sample taken at COGEMAC

4.3 Printing Settings

The marble powder printing parameters to perform a civil engineering constructability test depend on:

- Density of marble powder: between 1.5 and 2.5 g/cm^3
- Print speed: between 20 and 40 mm/s
- Layer thickness: between 1 and 5 cm

- Nozzle height: between 0.1 and 0.5 mm
- Filling percentage: between 10% and 30%
- Nozzle type: hardened steel or tungsten carbide nozzle can be used for marble powder printing. The nozzles used for 3D printing in construction have a diameter between 0.5 and 3 cm.

An example of the print settings for performing constructability tests, offered by Chen et al. (2021) [17], is presented in Table 2.

Table 2. Print settings for performing a constructability test [17]

Different dosages of binders	B45-1	B75-1	B75-2	B90-1	B90-2
Test duration (material age) [min]	15–30	15–30	15–30	15–30	15–30
Preset nozzle distance [mm]	8	8	8	8	8
Pump [rpm]	58–60	58–60	58–60	58–60	58–60
Material flow [L/min]	0.90	0.89	1.01	0.73	0.85
Nozzle opening (diameter) [mm]	15	15	15	15	15
Nozzle travel speed [mm/min]	5100	5040	5700	4140	4800
Time interval between layers [s]	7.4	7.6	6.7	9.0	8.1

4.4 Formulation of a Material for Use in 3D Printing in Civil Engineering

Mixing formulations of marble powder with other additives for 3D printing in civil engineering may vary depending on the properties desired for the final material. Some examples of possible formulations are:

- 80% marble powder and 20% gypsum
- 70% marble powder and 30% gypsum
- 60% marble powder and 40% gypsum
- 70% marble powder + 30% lime
- 60% marble powder + 40% lime
- 50% marble powder + 50% lime

These formulations can be adjusted according to the results of the characterization tests and the performance objectives of the final material. It is important to note that the quality and grain size of the marble powder and lime used can also influence the optimal formulation.

4.5 Mixing Element Testing after 3D Printing

The tests on the marble powder mixing element after 3D printing can include several parameters to assess the quality of the printed element. These parameters include:

- Compressive strength

- Bending strength
- Porosity
- Computed tomography (CT)
- Scanning electron microscopy (SEM).
- water absorption
- Surface roughness
- Dimensional Accuracy

5 Conclusion

This paper presents the potentialities of the valorization of Tunisian marble powder as a basic material for 3D printing in Civil Engineering. Mixed with other traditional local materials such as Lime and Gypsum, marble powder represents a potential candidate with the physico-chemical and mechanical characteristics to meet the requirements of printable materials. In this paper, an experimental 3D printing protocol and a formulation of the printable material have been proposed. The formulation process and the conduct of additive manufacturing tests were also presented.

References

1. Le, T.T., et al.: Hardened properties of high-performance printing concrete. Cem. Concr. Res. **42**(3), 558–566 (2012)
2. Labonnote, N., Rønnquist, A., Manum, B., Rüther, P.: Additive construction: state-of-the-art, challenges and opportunities. Autom. Constr. **72**, 347–366 (2016). https://doi.org/10.1016/j.autcon.2016.08.026
3. Nerella, V.N., Krause, M., Mechtcherine, V.: Practice-oriented buildability criteria for developing 3D-printable concretes in the context of digital construction (2018). https://doi.org/10.20944/preprints201808.0441.v1
4. Rahul, A.V., Santhanam, M., Meena, H., Ghani, Z.: 3D printable concrete: mixture design and test methods. Cem. Concr. Compos. **97**, 13–23 (2019)
5. Wangler, T., Roussel, N., Bos, F.P., Salet, T.A.M., Flatt, R.J.: Digital concrete: a review. Cem. Concr. Res. **123**, 105780 (2019)
6. Nair, S.A.O., Alghamdi, H., Arora, A., Mehdipour, I., Sant, G., Neithalathh, N.: Linking fresh paste microstructure, rheology and extrusion characteristics of cementitious binders for 3D printing. J. Am. Ceram. Soc. **102**(7), 3951–3964 (2019)
7. Wolfs, R.J.M., Bos, F.P., Salet, T.A.M.: Hardened properties of 3D printed concrete: the influence of process parameters on interlayer adhesion. Cem. Concr. Res. **119**, 132–140 (2019)
8. Wolfs, R.J.M., Bos, F.P., Salet, T.A.M.: Triaxial compression testing on early age concrete for numerical analysis of 3D concrete printing. Cem. Concr. Compos. **104**, 103344 (2019)
9. Wolfs, R.J.M., Bos, F.P., Salet, T.A.M.: Early age mechanical behaviour of 3D printed concrete: numerical modelling and experimental testing. Cem. Concr. Res. **106**, 103–116 (2018)
10. Wolfs, R.J.M., Bos, F.P., Salet, T.A.M.: Correlation between destructive compression tests and non-destructive ultrasonic measurements on early age 3D printed concrete. Constr. Build. Mater. **181**, 447–454 (2018)

11. Markin, V., Nerella, V.N., Schröfl, C., Guseynova, G., Mechtcherine, V.: Material design and performance evaluation of foam concrete for digital fabrication. Materials **12**(15), 2433 (2019). https://doi.org/10.3390/ma12152433
12. Neithalath lab (2019). 3-D Printing of Cementitious Binders| Neithalath Lab (2019). https://neithalath.engineering.asu.edu/3-d-printing-of-cementitious-binders/. Accessed 16 Aug 2019
13. Rahul, A.V., Santhanam, M., Meena, H., Ghani, Z.: Mechanical characterization of 3D printable concrete. Constr. Build. Mater. **227**, 116710 (2019)
14. Panda, B., Lim, J.H., Tan, M.J.: Mechanical properties and deformation behaviour of early age concrete in the context of digital construction. Compos. Part B Eng. **165**, 563–571 (2019)
15. Weng, Y., et al.: Feasibility study on sustainable magnesium potassium phosphate cement paste for 3D printing. Constr. Build. Mater. **221**, 595–603 (2019). **(Elsevier Ltd.)**
16. Benavides, A.: Northwestern researchers compete in NASA Mars challenge. The Daily Northwestern (2019). https://dailynorthwestern.com/2019/01/27/campus/northwestern-researchers-build-toward-3dprintable-habitat-on-mars. Accessed 16 Aug. 2019]
17. Chen, Y., He, S., Wan, Z., Çopuroglu, O., Schlangen, E.: 3D printing of calcined clay-limestone-based cementitious materials. Cem. Concr. Res. **149**, 106553 (2021). https://doi.org/10.1016/j.cemconres.2021.106553

Easy Conversion of PET Bottles to Eco-Filament for 3D Printing and Process Characterization

Chokri Ben Aissa[1,2](✉), Abd El Hedi Gabsi[2], Safa Mathlouthi[2,3], and Abdelkarim Ghanem[1,4]

[1] Laboratory of Mechanics, Materials and Processes, University of Tunis, ENSIT, 5 Taha Hussein Street, Montfleury, 1008 Tunis, Tunisia
chokri.benaissa@ensit.u-tunis.tn

[2] Higher Institute of Technological Studies of Nabeul, Mrezga University, Campus - 8000 Nabeul, Tunisia

[3] Applied Mechanics and Engineering Laboratory (LMAI), University of Tunis El Manar, National Engineering School of Tunis (ENIT-Tunisia), Campus El Manar, Tunis, Tunisia

[4] Higher Institutes of Technological Studies of Rades, Al-Quds Street, Rades, Tunisia

Abstract. Due to the economic and ecological impact, material recycling technologies are increasingly privileged and preferred over to standard processes. In this article, we focus on the potential to recover polyethylene terephthalate (PET) from water and soft drink bottles that have been poorly stored, transported and used, as well as from bottles from different manufacturers. The process used is to convert the bottle into a 1.75 mm diameter filament without fragmentation or dissolving. It does not require any heavy or industrial equipment and can be easily made even by 3D printing enthusiasts. We then used an FDM (Fused Deposition Modeling) 3D printer to build the modeled shape using the resulting filament.

The results of the process were characterized (tensile strength and hardness) and compared with those of 3D printing with "polylactic acid" (PLA), the most commonly used material in this type of printing. It offers adequate properties for many applications (close mechanical properties: elasticity around 230 MPa, maximum mechanical strength around 29 MPa, hardness around 10 MPa).

Keywords: Additive Manufacturing · 3D printer · FDM · Recycled PET · Mechanical characteristics

1 Introduction

The recycling of materials has become increasingly important in recent decades. The world produces millions of tons of plastic every year. Polyethylene terephthalate, also known as PET or PETE, is a thermoplastic polymer used to make a variety of goods. This substance is a recyclable thermoplastic composed of 62% carbon, 4% hydrogen, and 33% oxygen (Nisticò 2020). It can be processed in a different ways into a variety of shapes and colors. It is formed by the reaction of ethylene glycol and terephthalic acid. PET is known for its high strength and durability. It is a popular material for applications such as packaging, textiles and electronics due to its low deformation and shrinkage rates and transparency.

T. Mabrouki et al. (Eds.): AIAM 2023, LNME, pp. 170–178, 2024.
https://doi.org/10.1007/978-3-031-47784-3_21

PET is also a reliable and adaptable material (Zander et al. 2019). Overall, PET is highly resistant to UV rays and chemical degradation, making it a reliable material for bottles and other industrial items.

In the 3D printing method, successive layers of material are built up to create a 3D part from a digital model. The technology has grown in popularity in recent years because of its ability to quickly create customized products without using traditional production processes. Thermoplastics, composites, metals and ceramics are some of the materials used in 3D printing. Material selection is influenced by the application and desired properties of the final product. Fused Deposition Modeling (FDM), Stereolithography (SLA), and Selective Laser Sintering (SLS), among others, are different types of 3D printing.

PET filament is available as PETG (polyethylene terephthalate), a material originally manufactured or recycled using conventional methods (Zander et al. 2019) (Navarro et al. 2008) (Gomes et al. 2022) (Jankauskaitė et al. 2008). PETG is attractive due to its low deformation and shrinkage, but it still has a high cost compared to recycled PET; Our new simple method is based on converting the strips cut from the bottle into filaments through a work-inspired mechanism (Tylman and Dzierzek 2020), designed and fabricated according to the design shown in Fig. 1.

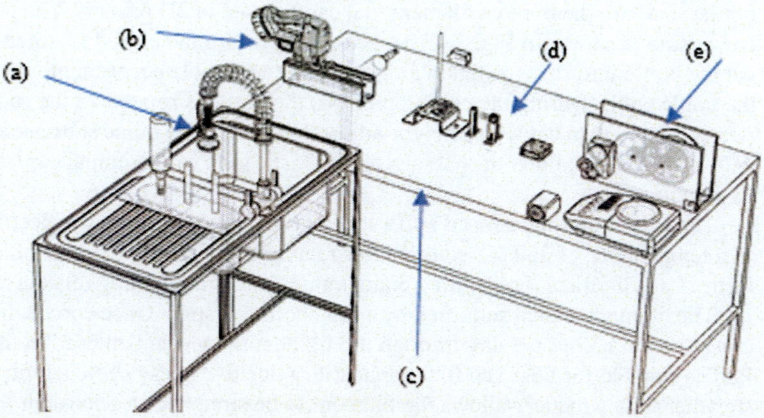

Fig. 1. Design of the mechanism for transforming bottles into filament; (a): bottle washing, (b): bottom cutout, (c): cutting into strips, (d): heating block and nozzle, (e): stretcher

Since the glass transition temperature of PET is relatively low (about 75 °C), it can be cooled and solidified immediately after extrusion. Therefore, the printed object will not warp during printing, creating a high-quality product. It is available in different colors and can be printed using different 3D printing techniques.

Given the accessibility of PET bottle waste, many cutting-edge technologies are currently being used to recycle it (Al-Sabagh et al. 2016). One such strategy is to recycle bottle leftovers and waste, which are then further processed into filaments for 3D printers. Recycling plastics for 3D printing is becoming popular solution due to increased environmental awareness and the need for sustainable production methods.

Extrusion is commonly used to produce polymer filaments for 3D printers. After being recycled and re-used, the PET material retains its properties (Exconde et al. 2019). As part of our technology, we manufacture 100% recycled filament from plastic bottles. It offers a less wasteful, simpler and more economical way to produce filament while reducing the environmental impact of waste plastic. This method produces uniform, fully recycled filaments (no virgin material being used) that can easily be fed into a 3D printer to create three-dimensional objects.

Therefore, the aim of this study was to compare the properties of PET recycled from bottles with polylactic acid (PLA), one of the most popular materials for 3D FDM printing, made from cornstarch.

Different types of filaments for 3D printing have been compared. However, for the studied materials (PET or PETG) (Roman et al. 2021), the filaments are manufactured using conventional techniques (extrusion).

2 Materials and Method

2.1 Elaboration

The first step in the project is to design and develop a mechanism that can recycle plastic bottles and turn them into a filament that can be used in 3D printers. The process of this conversion is shown in Fig. 2. First, blow hot air into the closed (washed) bottle, then cut the bottle into strips with a width of 9 mm (adjustable according to the thickness of the bottle wall). During the cutting process, the bottle is rotated on the rod while being stretched, and then cut with a height-adjustable blade. The guide subsystem is equipped with a groove that allows the passage of the plastic strip with minimal gaps, thus ensuring its stability.

The strip is then introduced into a special nozzle, heated with two electric resistances to a temperature of 250 °C, so that after relaxation, it emerges on the other side in the form of a cylindrical mold with a diameter of 1.75 mm - cooling relaxation.

The filament is then solidified by a fan cooling system. Once cooled, it is tested in a cross-section before passing through the tightening mechanism used in the 3D printer. PET is suitable for fiber and film applications due to its easy stretchability; therefore, a stretcher with a reducer allows the filament to be stretched at a constant low speed and wound into a spool for 3D printing.

In the second step, this filament is inserted into the 3D printer's extruder and the desired shape is created. To increase filament adhesion to the extruder drive and push the filament from both sides. Our custom 3D printers use a two-speed direct drive extruder. Also, use medium-diameter holes (0.5 mm) to prevent nozzle clogging.

The extruder's dual-gear design gives it a stronger grip on the filament than a single-gear extruder and allows it to apply more force to the filament as it extrudes.

This increased force allows the extruder to grab the filament and push it through the hot end, resulting in better and more efficient filament adhesion. Due to the direct drive design, the filament is pushed over a shorter distance. This fact reduces the space available for the filament to bend or kink, which also helps improve filament adhesion by preventing irregularities during extrusion.

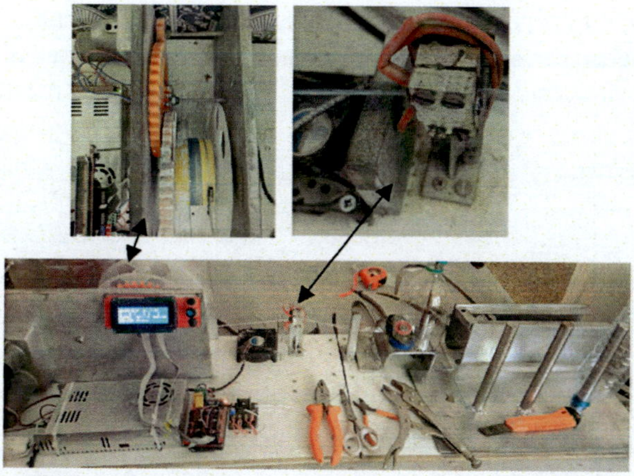

Fig. 2. Mechanism of converting bottles into filament

"PET printing" requires the printer to be heated to a nozzle temperature of 270 °C and a bed temperature of 70 °C compared to "PLA printing" which requires respectively a temperature of 220 °C and 50 °C. The translational speed during printing should not exceed 45mm/s to prevent PET from becoming amorphized. In this situation, a cooling fan is preferably used in case the melt cools, the molecules arrange themselves in an ordered and crystalline manner.

Each modeled sample was printed with two materials, 100% infill (lines), 1 mm thick walls and 0.2 mm thick deposition layer. Our customized printer is controlled by the programmable "Marlin" firmware on the Arduino Mega 2560 board and the slicer program is "Ultimaker-Cura". An example of a printed pattern is shown in Fig. 3.

(a) (b) (c)

Fig. 3. 3D printed specimens: (a) Calibration cube, (b) Printing of test specimens, (c) Tensile test specimens.

2.2 Characterization

Tensile tests were performed on Type B dumbbell specimens printed in PLA and PET material, in accordance with the ISO 3167 standard (Fig. 4). The tests were conducted using the "LLOYD LR 5K" test rig, which is depict-ed in Fig. 5. Three replicates of each test were performed, and the average value was calculated to determine the tensile strength.

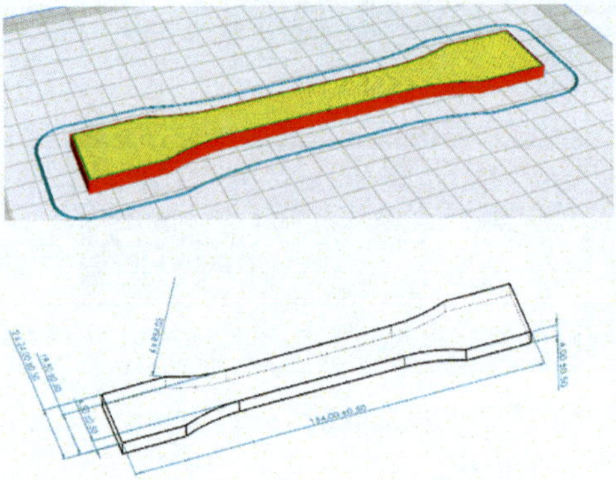

Fig. 4. Tensile specimen according to ISO 3167

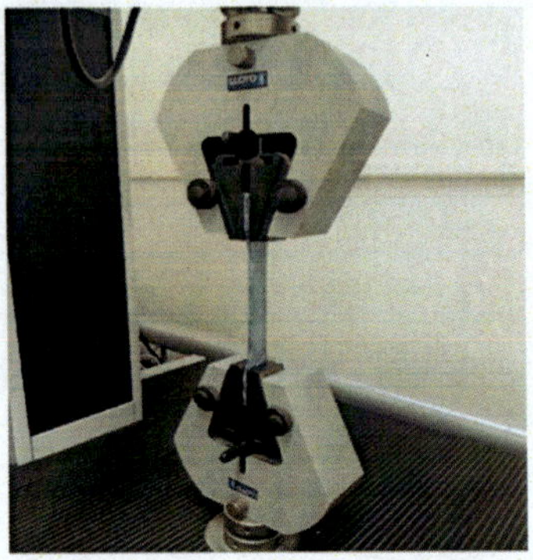

Fig. 5. Tensile test on PET and PLA specimens

Vickers hardness tests were performed on rectangular specimens (24 * 24 * 4 mm) made of PLA and PET using an instrumented INNOVATEST NEXUS series 7000 hardness tester (Fig. 6-*a*). This general-purpose tool features a square diamond grinding indenter with a 136° angle between the vertices between opposing faces.

Enter the diagonal values into the calculator using simple measurements, as shown in Figs. 6-*b* and 6-*c*, and read the indentation depth h immediately at the end of the indentation, as shown in Fig. 6-*d*. To determine the hardness value, each measurement is carried out three times and the average value is calculated.

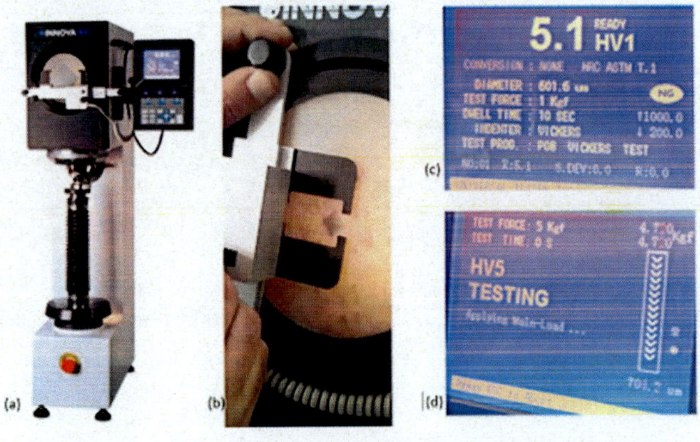

Fig. 6. Instrumented NEXUS type 7000 series hardness tester

The expression for Vickers hardness is:

$$HV = \frac{2\sin\frac{136}{2} * P}{(\frac{d_1+d_2}{2})2} = \frac{1.8544 * P}{(\frac{d_1+d_2}{2})2} \quad (1)$$

According to the experiments performed, PLA and PET materials are elastoplastic. Plasticity increases with increasing load.

The main difference between these curves is the relative importance of the remaining penetration depth (hc) to the maximum penetration depth (hmax) and the relative importance of elastic work (represented by the area under the curve when unloaded). Total work (expressed as the total area under the charging curve: $W_t = W_{pl} + W_{el}$). These two energies are shown in Fig. 7.

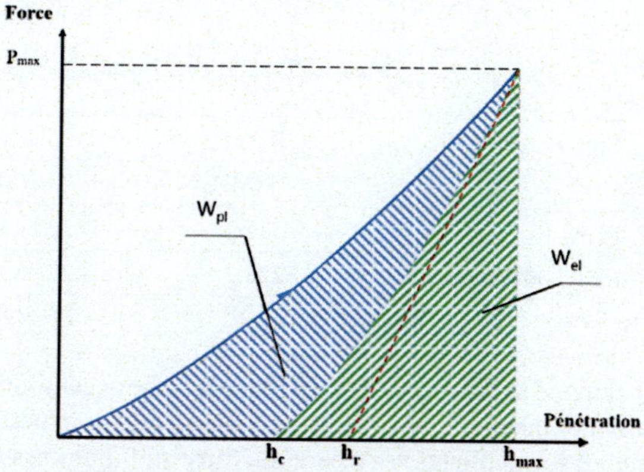

Fig. 7. Schematic representation of the plastic (W_{pl}) and elastic (W_{el}) work during loading-unloading in the hardness test

3 Results and Discussion

Tensile tests were conducted on specimens of the shape shown in Fig. 3. Initial tests revealed the importance of employing low velocities are necessary to avoid the formation of amorphous material caused by voids inside the sample, which would fracture under low stress. However, recent tests are encouraging as we have achieved very similar tensile strength values to industrial PLA (Values listed in Fig. 8 and Table 1).

The focus of this study was to confirm the potential of using 100% recycled PET as a base material in 3D printing and if it has at least the same properties as the PLA. Printing parameters can be further optimized to improve results.

First, PLA can withstand higher loads before failure, as its maximum load is slightly higher than recycled PET. For example, the final strength of recycled PET is only 28.35 MPa compared to 29.39 MPa for PLA. The elastic moduli of PLA and PET are typically 232.3 MPa and 228.7 MPa, respectively. However, both materials exhibit

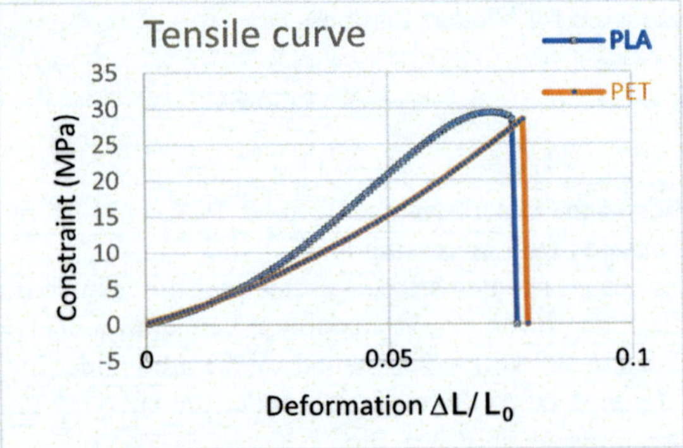

Fig. 8. Stress-strain curves of PLA and PET

Table 1. Tensile Test Results

Printed material	Results	
	E (MPa)	Rm (MPa)
PLA	232.3	29.39
PET 100% recycled	228.7	28.35

nonlinear behavior at higher strains. In this case, the secant modulus can be used as an approximation of the modulus of elasticity. Based on this experience, recycled PET may be better suited for applications where ductility and toughness are critical, while PLA may be better suited for applications requiring high strength.

When it comes to 3D printing, choosing the right material is critical to the quality, performance and cost of the final product. In terms of elasticity and ductility, PLA and PET have similar properties. This means that PET can effectively replace PLA for parts that do not require high opacity, while providing cost and environmental benefits as a recycled material. Also, PET showed better results with post-print shrinkage, which can be an issue with PLA and cause parts to separate from the print bed during the build process. PET printed parts also adhere better and are less prone to warping. 3D printing using PET instead of PLA is a viable option for manufacturing parts that require good mechanical properties, especially in terms of toughness and impact resistance. PET's low cost and recyclability make it an attractive alternative, especially for applications that do not require high opacity.

The stiffness of printed recycled PET depends on several factors, but with the right printing parameters, it is possible to obtain printed parts that are stiff enough for many applications. The results of the hardness test are represented in Table 2.

Unfortunately, this rating has its limitations because we are dealing with elastic materials whose hardness cannot be accurately assessed with available tools, so it is only a comparative indicator.

Table 2. Hardness Test Results

Printed material	Hardness (MPa)
PLA	7.34
PET 100% recycled	13.68

It's important to remember that when choosing a material, there may be other factors to consider when making the final decision, such as cost, availability, and ease of fabrication with a 3D printer.

Due to environmental concerns and increasing demand for sustainable practices, the use of recycled materials in industrial processes is becoming increasingly important. According to the findings, using recycled PET as a 3D printing material has several advantages, including lower costs and less negative environmental impact.

Using recycled PET reduces the environmental damage caused by plastic bottles. Given the important quantities of plastic waste that exist and can take hundreds of years to decompose, the impact could be significant.

4 Conclusion

This study develops a method for converting PET bottles into ecological filaments. The mechanical properties of PET and PLA were compared after printing (PLA is the most commonly used material in 3D printing).

According to our findings, the mechanical properties of 100% recycled PET are similar to those of PLA. This indicates that recycled PET can serve as a substitute for 3D printed parts without any compromise in performance.

According to our research, PET is harder and more malleable than PLA, which allows it to withstand more impact and stress before breaking. In contrast, the PLA is better suited for rigid parts that need to resist deformation under load because it is stiffer and stronger than PET.

Based on the results, it can be concluded that recycled PET is a more fa-vorable option for 3D printing compared to PLA, considering factors such as cost and environmental impact. Using recycled PET reduces the overall cost of the material while reducing the quantity of waste that ends up in landfills and minimizes the energy consumption associated with PLA production. Additionally, recycled PET offers a more economical and environmentally friendly solution than PLA when used as a 3D printing material.

To fully exploit the potential benefits of recycled materials, more research is needed to investigate the long-term consequences of using recycled PET in various applications and to improve 3D printing processes. Sources and recycling methods can have an impact on the quality of recycled PET, resulting in changes in the material's mechanical properties. Therefore, it is important to confirm that recycled PET complies with the specific standards for the application.

References

Al-Sabagh, A.M., Yehia, F.Z., Eshaq, Gh., Rabie, A.M., ElMetwally, A.E.: Greener routes for recycling of polyethylene terephthalate. Egypt. J. Petrol. **25**(1), 53–64 (2016). https://doi.org/10.1016/j.ejpe.2015.03.001

Exconde, M.K.J.E., Co, J.A.A., Manapat, J.Z., Magdaluyo, E.R.: Materials selection of 3D printing filament and utilization of recycled polyethylene terephthalate (PET) in a redesigned breadboard. Procedia CIRP **84**, 28–32 (2019). https://doi.org/10.1016/j.procir.2019.04.337

Gomes, T.E.P., Cadete, M.S., Dias-de-Oliveira, J., Neto, V.: Controlling the properties of parts 3D printed from recycled thermoplastics: a review of current practices. Polymer Degrad. Stab. **196**, 109850 (2022). https://doi.org/10.1016/j.polymdegradstab.2022.109850

Jankauskaitė, V., Macijauskas, G., Lygaitis, R.: Polyethylene terephthalate waste recycling and application possibilities: a review. In: MEDŽIAGOTYRA, vol. 14, Issue 2 (2008)

Navarro, R., Ferrándiz, S., López, J., Seguí, V.J.: The influence of polyethylene in the mechanical recycling of polyethylene terephtalate. J. Mater. Process. Technol. **195**(1–3), 110–116 (2008). https://doi.org/10.1016/j.jmatprotec.2007.04.126

Nisticò, R.: Polyethylene terephthalate (PET) in the packaging industry. Polym. Test.. Test. **90**, 106707 (2020). https://doi.org/10.1016/j.polymertesting.2020.106707

Roman, N., et al.: Materials for respiratory masks in the context of COVID 19 pandemic. Mater. Plastice **57**(4), 236–247 (2021). https://doi.org/10.37358/MP.20.4.5423

Tylman, I., Dzierzek, K.: Filament for a 3D printer from pet bottles- simple machine. Int. J. Mech. Eng. Robot. Res. **9**, 1386–1392 (2020). https://doi.org/10.18178/ijmerr.9.10.1386-1392

Zander, N.E., Gillan, M., Burckhard, Z., Gardea, F.: Recycled polypropylene blends as novel 3D printing materials. Addit. Manuf.. Manuf. **25**, 122–130 (2019). https://doi.org/10.1016/J.ADDMA.2018.11.009

Modeling and Optimization of Joint Condition Based Quality/Maintenance Strategy for an Additive Manufacturing Equipment

Zeineb Boumallessa[1](✉), Mounir Elleuch[2], Houssam Chouikhi[3], and Hatem Bentaher[2]

[1] Laboratory of Electromechanical Systems (LASEM), National Engineering School of Sfax, Sfax, Tunisia
Zeineb.boumallessa@gmail.com
[2] Higher Institute of Industrial Management of Sfax (ISGI), Sfax, Tunisia
[3] Mechanical Engineering Department, College of Engineering, King Faisal University, Al Hofuf, Kingdom of Saudi Arabia

Abstract. In industrial plants, operations management encompasses three main functions: quality management, production scheduling, and maintenance planning. While these functions are interconnected in reality, they are often treated separately in practice. However, it is widely recognized that handling these parameters in isolation can lead to suboptimal solutions. To address this issue, this paper proposes an integrated approach for jointly controlling production, quality, and maintenance in the context of additive manufacturing. Specifically, we focus on an Additive Manufacturing Equipment that produces a single type of product and undergoes a degradation process, resulting in the production of non-conforming units. To make informed decisions regarding the maintenance plan, inspections are conducted to monitor product quality and machine degradation. Our strategy involves establishing a relationship between the state of the quality indicator and the level of machine degradation, thereby studying their collective impact on the maintenance plan. The objective is to minimize the average total cost per unit of time, taking into account maintenance costs, quality-related costs, and production costs. To achieve this, we develop a mathematical model and provide a numerical example to calculate optimal values for the decision variables. The proposed approach is applied to address the maintenance planning challenge for 3D printer machines.

Keywords: Additive Manufacturing Equipment · Quality management · Production scheduling · Maintenance planning · Optimization

1 Introduction

Due to globalization and technological advancement, manufactories are faced to significant challenges. In recent years, additive manufacturing (AM) technology has gained significant recognition as a highly efficient method for producing intricate and complex

T. Mabrouki et al. (Eds.): AIAM 2023, LNME, pp. 179–188, 2024.
https://doi.org/10.1007/978-3-031-47784-3_22

objects across various industries. Its versatility and flexibility have opened up new possibilities in product design and manufacturing, allowing for the creation of components with intricate geometries and customized features. The wide range of applications of AM spans industries such as aerospace, automotive, healthcare, and consumer goods, to name a few. One of the key advantages of AM technology is its ability to eliminate many traditional manufacturing constraints. AM technology offers significant benefits in terms of cost and time efficiency. By eliminating the need for complex tooling and reducing material waste, AM can lead to cost savings in production. The ability to rapidly prototype and iterate designs also accelerates the product development process, reducing time-to-market and allowing for faster innovation cycles (Askari et al. 2020). The additive manufacturing process often requires less energy and produces less waste compared to traditional manufacturing methods. It also enables the use of recycled materials and the ability to create lightweight designs, contributing to resource efficiency and environmental conservation. As a result, the adoption of additive manufacturing has witnessed rapid growth across industries, with companies leveraging its advantages to gain a competitive edge. The continuous advancements in AM technology, materials, and process optimization further enhance its potential for revolutionizing the manufacturing landscape and driving innovation in product development (Khorasani et al. 2022). However, AM products should be manufactured with an advanced quality control process, which has the ability to prove and achieve the desired reliability and accuracy. Improving production performance has become a major challenge to maximize profits and maintain market advantages. Throughout the production cycles, the AM machines are susceptible to random breakdowns that can disrupt the production process, leading to possible interruptions. Maintenance is therefore essential to improve the availability of the system and the running of production, thanks to certain activities such as inspection, repair, replacement, etc. Quality control implies the improvement of product quality, which help to meet effectively the customer demand, and to reduce the costs of non-quality.

Managing quality, and maintenance has become increasingly complex due to various challenges. One major challenge arises from the frequent breakdowns that occur during the production cycle, leading to disruptions in production activities, increased downtime, a higher number of non-conforming products, and additional efforts and costs. As a result, companies are faced with the necessity of developing a robust production management policy that encompasses a carefully devised maintenance plan and an adequate quality control system. This integrated approach becomes imperative to ensure smooth operations, minimize downtime caused by breakdowns, optimize production efficiency, and uphold product quality standards. By implementing an effective production management policy, companies can mitigate the negative impacts of breakdowns, reduce the number of non-conforming products, and optimize overall production performance.

Recently, several research works proposed different joint approaches to reflect the interdependence between quality and maintenance in the Additive Manufacturing system. Many research articles have dealt with inspection and monitoring of AM procedures to improve the quality of printed physical objects. Recently, (Budzik et al. 2021; Charalampous et al. 2020; Liu et al. 2019) developed interesting work considering quality monitoring for metal-based additive manufacturing processes. The quality of a final AM product produced via a process has several mechanical properties, dimensional accuracy, surface quality, and conformance to standards and specifications. However, with traditional manufacturing processes, several tools and control methods have been developed. Some are still adaptable to additive manufacturing processes and their production.

But, the complexity of the parts sometimes makes these controls complicated or even impossible and, at the same time, new quality criteria are required with additive manufacturing. That's why, the tools and control methods must allow a complete inspection of the quality of the parts. To deal with quality control in in AM technology (Kim et al. 2018) implemented sampling plans to effectively monitor product quality. This integration allows to closely track the evolution of a quality indicator by employing control charts and statistical monitoring processes. This approach ensures that any deviations or issues in product quality are promptly identified and addressed, leading to improved overall performance. By combining control chart techniques and statistical monitoring, they manage maintenance activities and optimize the production system's performance. This integrated methodology not only enhances product quality but also contributes to cost savings and increased customer satisfaction. Researches adopted several types of quality control strategy (Gibson et al. 2021). Conventionally, these approaches have put forth a concept of expected cost of quality, which includes both the expenses incurred for inspections and the costs associated with non-conformity.

Based on the analysis performed and the trends of the listed papers, It seems clear that the integration of the quality control and maintenance plan of the Additive Manufacturing systems attracted the attention of researchers in recent years and several promising ways are to be developed. In this paper, we provide an improved methodology for joint consideration of maintenance plan and quality control based on the simultaneous monitoring of the degradation and the quality indicator and their both impact on the maintenance plan.

Being the contribution of this paper, presenting a methodology that allows to predict on additive manufacturing equipment preventive maintenance with the research aims to propose a new methodology to obtain the preventive maintenance principal elements at 3D machine printer.

The integrated model proposed in this study is innovative as it addresses the relationship between quality and maintenance by concurrently monitoring the quality indicator and the degradation of the AM system, and evaluating their impact on the maintenance plan. Unlike previous works on the same subject, the scenarios we have introduced have not been explored before. The remaining sections of the paper are organized as follows: Sect. 2 provides an overview of the industrial case and describes the problem statement along with the proposed strategy. In Sect. 3, we present an analytical study that includes the proposed mathematical model, followed by a numerical study that illustrates the proposed models. Finally, Sect. 4 concludes the paper and outlines potential avenues for future research.

2 Problem Statement

2.1 The Industrial Example Case

The plastic packaging industry includes companies that convert molten plastic into new packaging using 3D printing. In our study, we focus on the additive manufacturing process dedicated to 3D packaging for the creation of a protective case for an electronic component. The packaging manufacturing process includes several steps from customer ordering to shipping the final product. The considered machine is the additive manufacturing equipment. Its function is to produce individual packages from the melted plastic. To ensure that the product meets the quality requirements of customers, quality control using Xbar control chart is carried out at the end of the production process. Thus, an inspection of the state of deterioration of the AM equipment is carried out to verify the behavior of the machine.

The selection of the maintenance strategy depends on the parameters associated with product quality and system deterioration. The suitable maintenance approach can either be corrective maintenance or preventive maintenance.

In the packaging industry, companies are confronted with significant challenges in today's highly competitive industrial environment. The quality of the final product is directly influenced by the condition of the additive manufacturing (AM) machine. To address this issue and enhance product quality while reducing production costs, implementing an optimal maintenance plan becomes crucial. An effective maintenance program leads to improved system reliability and product quality.

In the following sections, we provide a comprehensive description of the problem we aim to tackle. By analyzing the relationship between maintenance activities, production processes, and product quality, we aim to develop a strategy that optimizes these aspects simultaneously.

2.2 Integrated Strategy

The aim of our study is to propose a novel approach that integrates maintenance and quality based on monitoring the AM equipment's degradation state and product quality control, considering various possible scenarios.

The model adopted in this study considers the case of a single AM machine producing a single product. We assume that the degradation of the AM machine affects the quality of the product (Surface quality). Therefore, based on the AM machine's degradation state and the level of product quality, decisions regarding the maintenance schedule and quality control will be made.

The Additive Manufacturing Equipment considered produces a single product at a constant rate p, the production should satisfy a constant and continuous demand d ($d < p$).

The detected failures affected the surface quality of the product, which increases the number of non-conforming products. Every h time unit, inspections are performed to control the AM degradation and the surface quality.

The main objective of this integrated strategy is to determine the optimal parameters of the control chart which minimizes the total cost of production.

The diagram in Fig. 1 summarizes the maintenance/ quality strategy and highlights the different possible states of integration between the AM system degradation and the surface of the product quality. The policies adopted reflected their simultaneous impact on the maintenance plan.

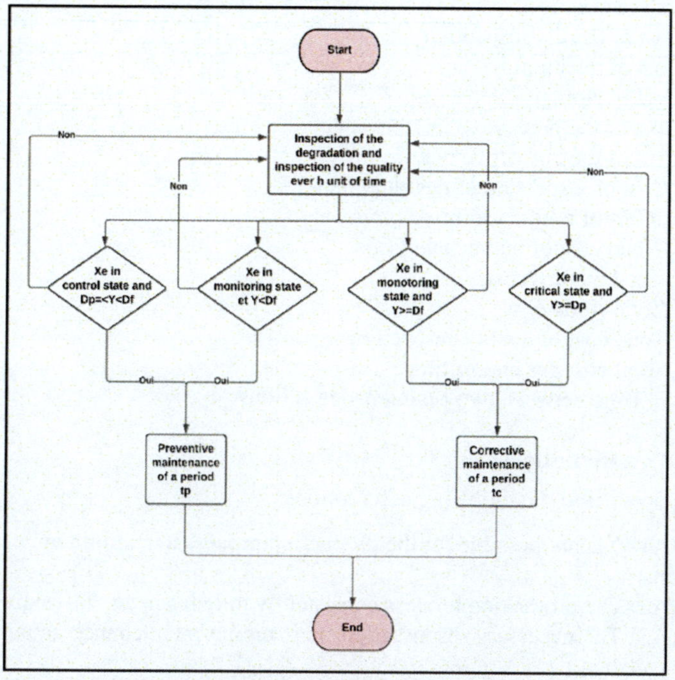

Fig. 1. Maintenance/Quality policy Flowchart

2.2.1 Notation

We propose the following notations:

p Production rate
d Demand rate
n Sample size
h Sampling interval
X_i Measurement quality indicator
Y Degradation level of AM.
k_s Deviations between the center line of the control chart and the surveil-lance limits
k_c Deviations between the center line of the control chart and the control limits
P_d Probability of occurrence of 1st scenario
P_p Probability of occurrence of 2nd scenario
P_f Probability of occurrence of 3rd scenario

P_c Probability of occurrence of 4th scenario
P_{in} Probability of occurrence of the 'in-control' state.
t_p Preventive maintenance duration
t_c Corrective maintenance duration
D_p PM threshold
D_f failure threshold
C_i Cost of quality inspection
C_r Cost of rejection
C_s Cost of storage
C_p Cost of the preventive maintenance
C_c Cost of the corrective maintenance
C_{im} Cost of inspection of degradation
EC_{nc} Cost of non-conforming products
EC_m Total cost of maintenance cost
EC_q Total quality cost
EC_s Total stock cost
EC_i Total cost of quality-inspection
EC_t Total cost per unit of time
D_{cycle} Time between two Maintenance actions

2.2.2 Assumptions

We considered the following assumptions:

- (A1) It is not possible for the production process to exhibit two scenarios simultaneously.
- (A2) The production process consistently initiates in an "in-control" state.
- (A3) The necessary resources for performing maintenance actions are consistently available
- (A4) The non-conformity of the product is attributed to the degradation of the AM system; however, it is important to note that they are not directly proportional.
- (A5) Products that do not meet the required specifications are rejected.
- (A6) All costs associated with inventory, maintenance, and quality are predetermined and remain constant.
- (A7) The construction time of AM machines is presumed to be influenced by both the AM machine's building technique and the design of the part being produced.
- (A8) The degradation state of the system is determined through inspections, whereas failures are self-reported.

2.3 Mathematical Model

Understanding the performance of an AM system is crucial for ensuring its operation, achieving set objectives, and making decisions regarding its future development. It is necessary to select and utilize performance indicators that serve as benchmarks for interpreting and comparing various information related to the system's performance.

The considered cost indicators corresponding to the proposed AM system are as follows:

- Preventive maintenance cost: This is the total cost incurred for performing planned maintenance activities aimed at preventing surface state failures.
- Corrective maintenance cost: This is the total cost incurred for repairing failures in the AM system. This may include costs of spare parts, labor costs, etc.
- Non-quality cost: This is the total cost of non-conforming or defective products produced by AM. This may include costs of rejection, rework, or reprocessing of non-conforming products.
- Stock cost: This is the cost associated with maintaining a level of inventory to compensate for production fluctuations caused by system failures.
- Quality inspection cost: This is the total cost incurred for conducting inspection activities to detect surface quality defects and ensure product conformity.
- Degradation inspection cost: This is the total cost incurred for monitoring and assessing the degradation of the production system over time. This may include costs of regular inspections to detect signs of degradation, costs of monitoring performance parameters, costs of analyzing collected data, etc.

The first step in developing the joint strategy is to establish a mathematical model that captures the relationship between the condition of the AM equipment, product quality, and maintenance actions. This model should consider factors such as degradation rates, failure thresholds, and quality indicators. By incorporating these parameters, a holistic view of the system can be obtained, enabling effective decision-making for maintenance and quality control.

The goal is to identify the optimal parameters that minimize the average total cost per unit of time over a restoration cycle. These parameters include:

- The optimal sample size, denoted as n*
- The optimal sampling interval, represented by h*
- The coefficients for monitoring and control limits, denoted as ks* and k_c*.

To meet the demand d, the additive manufacturing (AM) system operates continuously, and its state follows one of the four proposed scenarios. During production, the system halts at a certain stock level, and either preventive or corrective maintenance actions are carried out. The average total cost is composed of the average maintenance cost and the average quality cost, which includes expenses for sampling and rejecting non-conforming products. The average quality cost includes the average cost of quality inspections and the average cost of non-conforming products. Additionally, the average total maintenance cost for a restoration cycle includes the costs of preventive and corrective maintenance, along with the expenses associated with degradation inspections.

The total cost per unit of time is calculated as follows:

$$EC_t = \frac{EC_s + EC_m + EC_q}{D_{cycle}} \tag{1}$$

$$EC_s = C_s \times \begin{Bmatrix} P_{in} \times (\mathrm{p}-\mathrm{d}) \times h\left(\frac{N_c}{2} + N_s + \frac{N_i}{2}\right) \\ +P_c \times \left(t_c \times ((\mathrm{p}-\mathrm{d}) - \alpha_c) - \frac{t_c{}^2 \times \mathrm{d}}{2}\right) \\ +P_p \times \left(t_p \times ((\mathrm{p}-\mathrm{d}) - \alpha_p) - \frac{t_p{}^2 \times \mathrm{d}}{2}\right) \\ +P_f \times \left(t_c \times ((\mathrm{p}-\mathrm{d}) - \alpha_p) - \frac{t_c{}^2 \times \mathrm{d}}{2}\right) \\ +P_d \times \left(t_p \times ((\mathrm{p}-\mathrm{d}) - \alpha_d) - \frac{t_p{}^2 \times \mathrm{d}}{2}\right) \end{Bmatrix} \tag{2}$$

$$\mathrm{EC_q} = \mathrm{EC_i} + \mathrm{EC_{nc}} \tag{3}$$

$$\mathrm{EC_m} = \mathrm{EC_{PM}} + \mathrm{EC_{CM}} \tag{4}$$

3 Numerical Study

This section presents a numerical example that aims to validate the effectiveness of the proposed approach. The AM system operates with a production rate of $\mathrm{p} = 100$ units per time, while the fixed demand is set at 40 units per time. The quality indicator is modeled as a random variable following a normal distribution with a mean of $\mu = 3$ and a standard deviation of $\sigma = 2$. On the other hand, the degradation process of the system is described by a stationary gamma stochastic process $\{\mathrm{g(t)}: \mathrm{t} \geq 0\}$, where the scale parameter is $\eta = 20$ and the shape parameter is $\gamma = 2$. The production system is considered to fail once its degradation reaches the threshold value of $\mathrm{D_f} = 10$ and is deemed unreliable when the degradation level is $\mathrm{D_p} = 8$.

We propose the bellow costs and the durations:

$$\mathrm{t_p} = 1,\ \mathrm{t_c} = 3,\ \mathrm{C_i} = 3,\ \mathrm{C_p} = 100,\ \mathrm{C_c} = 4000,\ \mathrm{C_r} = 50,\ \mathrm{C_{im}} = 20,\ \mathrm{Cs} = 1.$$

The optimization problem was tackled using Mathematica software to determine the optimal values of h*, n*, $\mathrm{k_s}$*, and $\mathrm{k_c}$* that minimize the average total cost.

Based on Table 1, the analysis revealed that for Experiment #1, the optimal total cost per unit of time was determined to be 152.29. This result was obtained for an optimal sample size of 285.131 samples and for an optimal sampling interval of 7 units of time. On the other hand, Experiment #2 showed the highest total cost per unit of time,

Table 1. Results generated for different experiments

Experiments	$\mathrm{P_p}$	$\mathrm{P_d}$	$\mathrm{P_f}$	$\mathrm{P_c}$	n*	h*	$\mathrm{K_s}$*	$\mathrm{K_c}$*	$\mathrm{EC_t}$
Exp1	0,1	0.0	0.0	0.0	285.131	7	1	4	152,29
Exp2	0.0	0,1	0.0	0.0	1653.2	41	2	3	204,23
Exp3	0.0	0.0	0,1	0.0	1367	34	2	3	174,16
Exp4	0.0	0.0	0.0	0,1	68.39	41	2	3	190

reaching 204.23. These experiments provided a basis for comparing the optimal cost values, highlighting that scenarios P_p and P_f incurred lower costs compared to the other scenarios.

The findings clearly indicate that implementing a predictive preventive maintenance policy leads to cost optimization. By utilizing the optimal parameters, organizations can effectively manage their maintenance plan, ensuring its efficiency and effectiveness. By minimizing maintenance-related costs and maximizing equipment uptime, companies can achieve significant cost savings and gain a competitive advantage in the market.

Implementing a joint condition-based quality and maintenance strategy brings several benefits to AM systems. Firstly, it enhances product quality by proactively addressing equipment degradation and identifying potential quality issues early on. Secondly, it optimizes maintenance efforts, reducing unnecessary downtime and associated costs. Finally, it enables resource allocation based on real-time condition monitoring, leading to efficient utilization of personnel, materials, and equipment.

4 Conclusion

The quality control methodology proposed in this paper is applicable to additive manufacturing products made from any material. The utilization of the Xbar control chart enables the evaluation of the manufacturing process by utilizing digital data obtained from computer systems. This approach aligns well with the principles of Industry 4.0, as it is compatible with modern manufacturing systems that embrace advanced technologies and digitalization.

To summarize, models produced using additive manufacturing can undergo diverse quality control processes depending on their specific application, and this factor should be considered when choosing appropriate quality assessment methods. Additionally, it is crucial to always take cost aspects into account, as not all products necessitate high levels of precision and durability.

This paper proposed a strategy that integrates maintenance, production, and quality control, for an Additive Manufacturing Equipment subject to deterioration that affects the product conformity. We analyzed four distinct scenarios by concurrently monitoring two parameters: the state of the quality indicator and the level of degradation. We examined how these parameters influenced the maintenance plan. Our approach involved establishing a connection between the quality indicator and the machine's degradation state through a coupling function. This function represented the interdependence between the two parameters. To identify the optimal values for the control chart parameters, we employed a developed mathematical model that facilitated striking the best balance among various problem parameters. These optimal values were determined to minimize the average total cost rate, encompassing expenses related to storage, inspections, maintenance, and expected quality.

By considering the degradation of the machine and its influence on product quality, the proposed model provides a comprehensive approach to managing maintenance, production, and quality control. It enables decision-making regarding maintenance actions and quality control measures to minimize the production of non-conforming products and optimize the overall system performance.

This model can be applied across various industries and sectors that utilize AM technology, allowing manufacturers to improve the reliability of their machines, enhance product quality, and reduce the costs associated with non-conformities.

References

Askari, M., et al.: Additive manufacturing of metamaterials: a review. Addit. Manuf. **36**, 101562 (2020). https://doi.org/10.1016/j.addma.2020.101562

Khorasani, M., et al.: A review of industry 4.0 and additive manufacturing synergy. Rapid Prototyp. J. **28**, 1462–1475 (2022). https://doi.org/10.1108/RPJ-08-2021-0194

Budzik, G., Woźniak, J., Paszkiewicz, A., Przeszłowski, Ł, Dziubek, T., Dębski, M.: Methodology for the quality control process of additive manufacturing products made of polymer materials. Materials **14**, 2202 (2021). https://doi.org/10.3390/ma14092202

Charalampous, P., Kostavelis, I., Tzovaras, D.: Non-destructive quality control methods in additive manufacturing: a survey. Rapid Prototyp. J. **26**, 777–790 (2020). https://doi.org/10.1108/RPJ-08-2019-0224

Liu, C., Law, A.C.C., Roberson, D., Kong(James), Z.: Image analysis-based closed loop quality control for additive manufacturing with fused filament fabrication. J. Manuf. Syst. **51**, 75–86 (2019). https://doi.org/10.1016/j.jmsy.2019.04.002

Kim, H., Lin, Y., Tseng, T.-L.B.: A review on quality control in additive manufacturing. Rapid Prototyp. J. **24**, 645–669 (2018). https://doi.org/10.1108/RPJ-03-2017-0048

Gibson, I., Rosen, D., Stucker, B., Khorasani, M.: Additive Manufacturing Technologies. Springer International Publishing, Cham (2021).https://doi.org/10.1007/978-3-030-56127-7

Experimental and Numerical Study of the Wire Arc Additive Manufacturing (WAAM) Process

Seddik Shiri[1,2,3](✉), Mounir Rabhi[2], Ated Ben KhalifA[2], Khaled Boulahem[3], Fethi Aloui[4], and Sami Chatti[2]

[1] Laboratory of Mechanics, Production, and Energy (LMPE), 1008 Tunis, Tunisia
seddik.shiri@enib.ucar.tn
[2] National Engineering School of Monastir, 5019 Monastir, Tunisia
[3] National Engineering School of Bizerte, 7035 Menzel Abderrahman, Tunisia
[4] Fondouk Jdid Military Academy, 8012 Nabeul, Tunisia

Abstract. Wire Arc Additive Manufacturing (WAAM) is an additive manufacturing process that utilizes an electric arc to melt and deposit metal wire layer by layer, creating a three-dimensional component. This process offers several advantages, such as the ability to produce large-sized parts. It is increasingly being utilized in industries for the production of components for aerospace, automotive, shipbuilding, and other applications. However, it also comes with drawbacks, including low surface resolution and the need to manage deformations and residual stresses. Continuous research and development in this technology are crucial to optimize manufacturing parameters and enhance the mechanical properties and quality of produced parts. In this study, an experimental investigation was conducted to determine the optimal parameters of the WAAM process for producing complex parts. Simple and complex components were fabricated using this process. A numerical model of the WAAM process was developed using the LS-DYNA code. This numerical simulation can aid in understanding the material behavior during the welding process.

Keywords: Additive Manufacturing · WAAM (Wire Arc Additive Manufacturing) · Experimental Study · Numerical Modeling

1 Introduction

Wire Arc Additive Manufacturing (WAAM) is an emerging 3D printing technology that has gained significant attention in recent years. It offers unprecedented opportunities in various industrial sectors, ranging from aerospace to tooling manufacturing. WAAM is a 3D printing technique that utilizes an electric arc to melt a metal wire and deposit successive layers to create a three-dimensional component.

While WAAM holds great potential for cost-effective and large-scale production of complex metal parts, it faces several technical challenges. First and foremost, ensuring the quality of WAAM-fabricated parts is crucial to guarantee their structural integrity and required mechanical properties. This necessitates a profound understanding of the

T. Mabrouki et al. (Eds.): AIAM 2023, LNME, pp. 189–200, 2024.
https://doi.org/10.1007/978-3-031-47784-3_23

underlying physical phenomena occurring during the printing process and the ability to precisely control the manufacturing parameters.

To address these challenges, extensive research has been conducted in recent years to enhance the WAAM process (Gowthaman et al. 2023, Zhao et al. 2023, Gurmesa and Lemu 2023, Kumar et al. 2022, Shah et al. 2023). These studies have focused on various aspects, such as optimizing the manufacturing parameters, modeling and simulating the thermal and mechanical phenomena, and developing novel quality control strategies. For instance, (Gowthaman et al. 2023) investigated the effect of welding parameters on the microstructure and mechanical properties of WAAM-fabricated parts. The results of their study underscored the importance of appropriate parameter selection to achieve high-quality parts. On the other hand, (Kumar et al. 2022) proposed a numerical modeling approach based on the finite element method to simulate the thermal and mechanical phenomena during the WAAM process.

Within the scope of this study, we present an integrated approach comprising experimental investigations and numerical simulations to gain deeper insights into and improve the WAAM process. The first part of this paper focuses on the experimental aspect, where we conduct manufacturing trials of metal parts using WAAM and perform mechanical analyses to evaluate the quality of the produced parts. In the second part, we develop a numerical model based on the finite element method to simulate the thermal and mechanical phenomena during the printing process.

2 Experimental Study

During this work, we will try to improve manufacturing parameters, materials, environmental conditions and the various variables that can affect the quality of the manufactured part. To this end, we begin this section by presenting the experimental procedures for producing WAAM parts.

2.1 Equipment

In this study, a WAAM system based on a Router Milling ZX-4030 CNC machine was used, as shown in Fig. 1. The equipment was configured and customized for the WAAM process. This included mounting the wire spool, preparing the part to be manufactured, adjusting the power source parameters and installing the welding unit.

The welding torch is mounted to a 3-axis controlled workstation. The TPs 320i was used as a deposition heat source via a cold metal transfer (CMT) modulus that enables deposition with low heat input compared to other arc heat sources. The CNC 3-axis machine is coupled with an interface communication with the CMT power source TPS 320i.

2.2 Materials

The filler metal (FM) used for welding is a non-alloyed steel wire F57 of the brand SELECTARC (Xometry 2015). A S235 metal plate of size 120 mm × 120 mm x 5 mm was used as the substrate plate for deposition. The ends of the substrate were fixed by a holding device along the XY-axis table.

The shielding gas used for all tests was a mixture of argon (80%) and carbon dioxide (20%). The welding parameters are shown in Table 1 below. The choice of these parameters is based on a preliminary study involving the choice of material, welding electrical parameters (Table 2) and displacement speed (Table 3) to assess the optimum parameters for obtaining good weld quality.

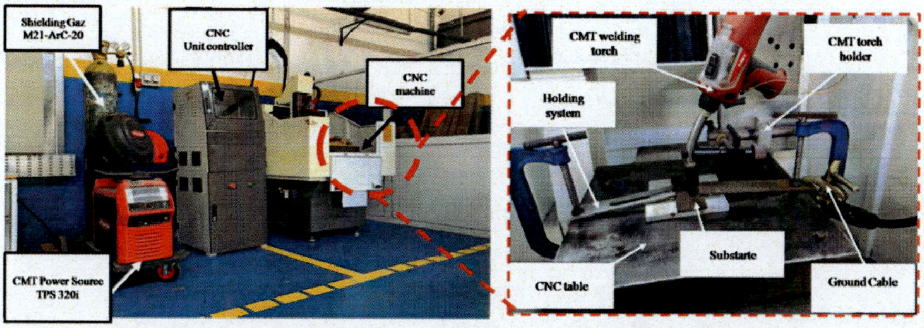

Fig. 1. WAAM experimental set-up: red lines indicate zoom.

The shielding gas used for all tests was a mixture of argon (80%) and carbon dioxide (20%). The welding parameters are shown in Table 1 below. The choice of these parameters is based on a preliminary study involving the choice of material, welding electrical parameters (Table 2) and displacement speed (Table 3) to assess the optimum parameters for obtaining good weld quality.

The deposition speed of the metal wire plays a crucial role in WAAM and can significantly impact the quality of the produced parts. It is a critical parameter that needs to be carefully controlled and optimized during the manufacturing process.

When the deposition speed is too high, it can result in surface defects on the printed part. The excessive speed may not allow sufficient heat transfer and fusion between the deposited layers, leading to incomplete bonding and inadequate interlayer adhesion. This can manifest as surface irregularities, such as roughness, waviness, or even discontinuities, compromising the overall integrity and functionality of the part.

Conversely, when the deposition speed is too low, it can lead to fusion defects and porosity. In this case, the heat input may be excessive due to slower wire deposition, causing prolonged exposure to the heat source. This prolonged exposure can result in excessive melting and overheating, leading to issues like excessive grain growth, solidification cracking, or the formation of voids and porosity within the printed part. These defects can significantly weaken the structural integrity and mechanical properties of the component.

Optimizing the deposition speed in WAAM is crucial to achieve a balance between the material deposition rate and the heat input. It involves finding the optimal speed that ensures sufficient heat transfer for complete fusion and bonding between layers while avoiding excessive heat input that can lead to defects. This optimization process often involves iterative testing and experimentation, along with advanced monitoring techniques to assess the quality of the printed parts.

Melting parameters, such as electrical voltage and arc intensity, can influence the quality of the part produced. Wrongly chosen melting parameters can lead to defects such as cracks and porosity, while well-chosen melting parameters can improve melting quality.

Table 1. Welding parameters.

Welding parameters	Value
Voltage: U (V)	10,4
Current: I (A)	87
Wire feed speed: WFS (m/min)	2.5
Travel speed: Ts (mm/min)	400
Waiting time: Tr (s)	30
Distance between beads: D (mm)	1.5
Tube contact distance (mm)	15

Table 2. Average values for current, voltage, power and associated wire melting speed.

		P1	P2	P3	P4	P5	P6	P7	P8	P9	P10
		Average values									
Current	A	40	47	51	71	87	104	138	155,6	185	205
Voltage	V	10,1	10,1	10,1	10,4	10,4	10,4	10,4	12,5	13,2	14,26
Power	W	404	475	515	738	905	1082	1445	2219	2312	2706
Wire fusion speed	m/min	1,1	1,2	1,4	2	2,5	3	4	5	6,3	7,8
Bead geometry	-	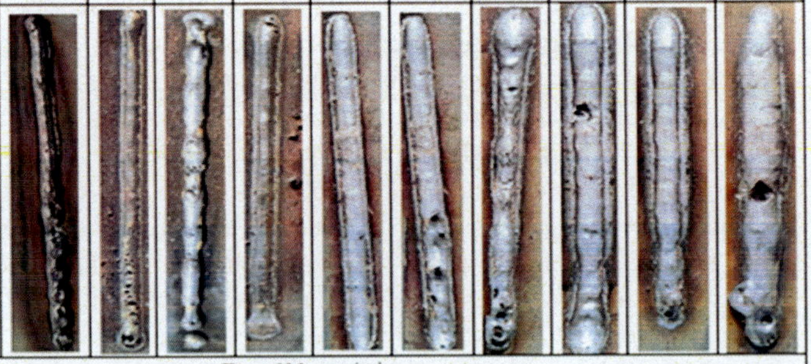									
Ts = 600 mm/min											

Table 3. Height and width of welded beads obtained for different travel speeds.

Travel speed (mm/min)	welded beads	Height and widht (mm/min)
300	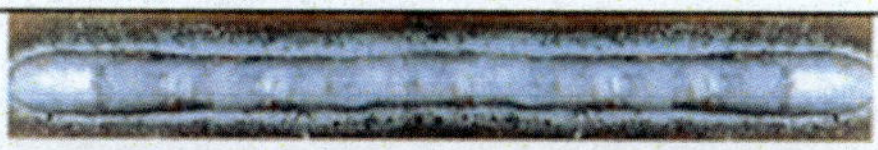	h=2.5 w=2.3
600		h=2 w=2
900	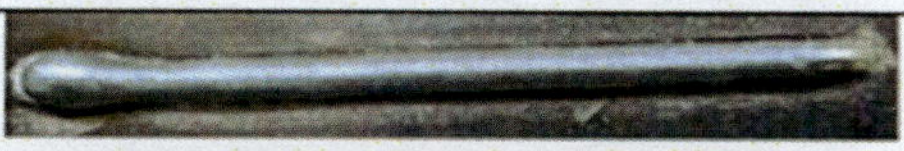	h=1.5 w=1.2
1200		h=1.2 w=0.8
1500		h=0.9 w=0.4

2.3 Methods

In the WAAM process, there are several types of material deposition path: spiral path, zigzag path, helix path and contour path. The material deposition path can also affect the quality of the part produced. A poorly chosen trajectory can lead to surface defects and excessive roughness, while a well-designed trajectory can improve surface quality and reduce porosity. Figure 2 shows the effect of two manufacturing strategies: Zig-Zag and unidirectional, on the quality of the part produced. The latter strategy shows the defect of material collapse.

The right choice of material deposition strategy optimizes surface quality and reduces defects such as porosity and cracks. The optimum parameters found in a previous study enabled us to produce complex shapes as shown in Fig. 3 by adopting the following optimum parameters: travel speed 900 mm/min; layer height 1.5 mm; pause time between two layers 30 s and an alternating deposition strategy.

3 Numerical Modeling

The Wire Arc Additive Manufacturing (WAAM) process is a promising technique for the fabrication of complex metallic structures. However, due to the complexity of thermomechanical interactions during the manufacturing process, it is crucial to develop a thorough understanding of the behavior of structures manufactured using WAAM. In this context, numerical modeling plays a crucial role in analyzing and predicting the behavior of parts produced by this process.

The objective of this study is to develop a numerical model capable of accurately representing the different stages of the WAAM process, including wire fusion, cooling, and springback. By taking into account the actual process parameters, our model

Travel speed (mm/min)	welded beads	Height and widht (mm/min)
300		h=2.5 w=2.3
600		h=2 w=2
900		h=1.5 w=1.2
1200		h=1.2 w=0.8
1500		h=0.9 w=0.4

(a) (b)

Fig. 2. Effect of manufacturing strategy: (a) Unidirectional strategy; (b) Zig-Zag strategy.

Fig. 3. Complex shapes achieved with optimal welding parameters.

will provide reliable results and enhance our understanding of the thermomechanical mechanisms that influence the quality and stability of fabricated structures.

To develop our numerical model, we rely on advanced thermomechanical equations that consider heat transfer, plastic and elastic deformations, as well as solidification phenomena. We have also incorporated the specific thermal and mechanical properties of the material used in the WAAM process.

In this section, we will present our numerical model in detail, explaining the assumptions and approximations made, as well as the equations and numerical methods used to solve the complex thermomechanical problems encountered during the WAAM process. We will also discuss the results obtained through our model.

3.1 Geometrical Specifications of the FEM Model

The FE software LS-DYNA was used to build a multi-layer model using thermomechanical analysis. Two additive beads having the same dimensions (3 mm × 5 mm

× 600 mm) were deposited on a base plate (9 mm × 500 mm × 600 mm) (see Fig. 4a). The substrate plate is constrained according to Fig. 4b.

Fig. 4. a) Parts geometry modeling. b) Substrate plate constraints.

The FE-model consists of 29921 solid hexahedron elements. The mesh has been refined around the weld area. A mesh with three elements in the height and four elements in the width of the weld bead was used (see Fig. 5a).

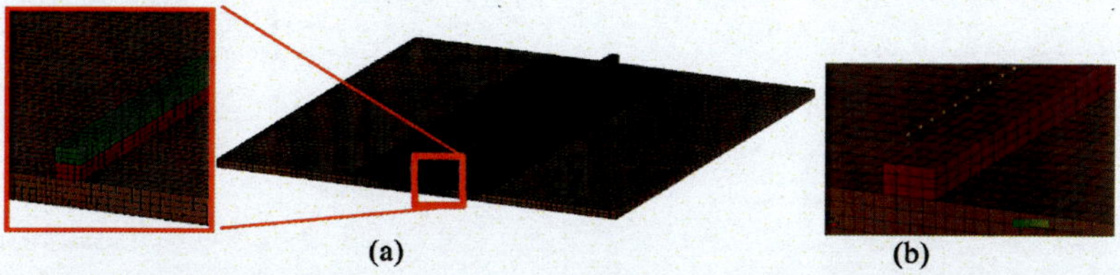

Fig. 5. (a) Meshing of the FE model. (b) Trajectory of the root of the weld torch defined by a node set.

A uni-directional path strategy was used to stack the two layers onto each other. The LS-DYNA keyword *BOUNDARY_THERMAL_WELD_TRAJECTORY was used to simulate a weld torch (beam element) moving along a path defined by a node set according to Fig. 5b. The mechanical tied surface to surface contact (*CONTACT_TIED_SURFACE_TO_SURFACE) was used to connect the first bead to base plate. The two welds were connected by thermal contact activated during the process.

3.2 Material Properties and Weld Heat Source Model

The numerical model takes into account the initial temperature of the metal plates, convection, and radiation. The thermal analysis was conducted using the *MAT_Thermal_CWM function in LS-DYNA. The FE model elements can be classified as "solid," "liquid" or "Ghost."

When a material is defined as "solid," it signifies that it has been activated in a previous welding pass. This indicates that the material has undergone a previous welding process and possesses the corresponding thermal and mechanical properties. Conversely, when a material is defined as "liquid," it means that it will be activated in the current welding pass. This indicates that the material will undergo a fusion process and become liquid as the welding heat reaches a specified temperature. On the other hand, when a material is defined as "Ghost," it means that it will not be activated in the current welding pass.

This can be used to model situations where a specific material does not participate in the welding process at a given stage.

Material activation occurs when the temperature reaches a specified value. At this stage, the material is activated and assumes the properties of the specified material, such as thermal conductivity, specific heat capacity, and thermal expansion coefficients. This enables realistic modeling of phase changes in the material and associated thermal effects.

The properties of the temperature dependent material were taken from (Israr et al. 2018) and given in Fig. 6.

The thermo-elastic-plastic model *MAT_CWM in LS-DYNA is used. This model uses the true stress-strain curves at different temperatures taken from (Mikael et al. 2016) and given in Fig. 7.

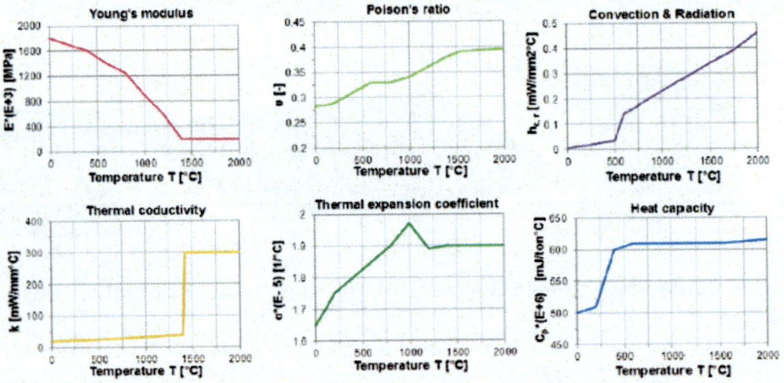

Fig. 6. Mechanical and thermal material properties for steel from (Israr et al. 2018).

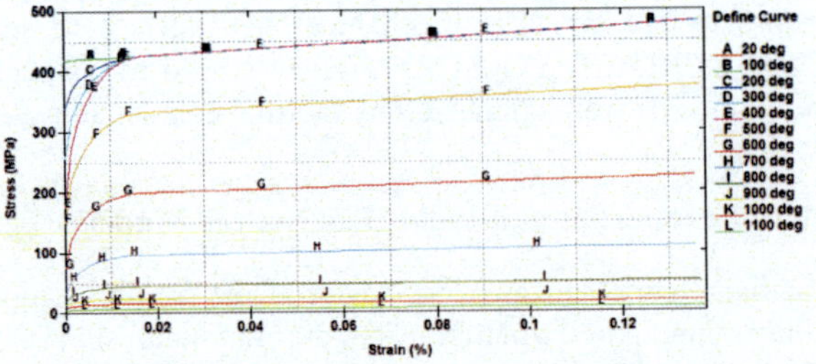

Fig. 7. True stress-strain curves at temperatures in the range of 20–1100 °C (Mikael et al. 2016).

The weld source is modeled using a Goldak double ellipsoid heat source given by Eq. 1.

$$q = \frac{2n\sqrt{n}FQ}{\pi\sqrt{\pi}abc}, \exp\left(\frac{-nx^2}{a^2}\right) \exp\left(\frac{-ny^2}{b^2}\right) \exp\left(\frac{-nz^2}{c^2}\right) \quad (1)$$

$$\text{The weld pool energy is expressed by} : Q = \eta UI \quad (2)$$

This model is commonly used in the fields of thermal engineering, metallurgy, welding process simulation, and heat modeling. The model is based on the geometric shape of two superimposed ellipsoids, which represent the heat source (see Fig. 8). By using thermal equations and heat transfer principles, the Goldak model allows for the calculation of heat distribution within a material or structure.

Based on input parameters such as the power of the heat source, position, and orientation of the ellipsoids, the Goldak model can predict the distribution of temperature and heat flux in the studied area. This enables understanding of how heat propagates and dissipates, as well as evaluating the thermal effects on materials.

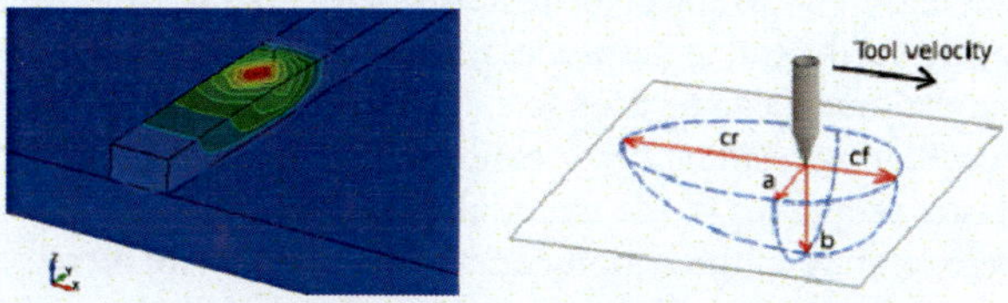

Fig. 8. Goldak's heat source distribution pattern.

The Goldak's ellipsoidal parameters used in this work are chosen in accordance with [8] and their corresponding values are given in Table 4.

Table 4. Goldak's ellipsoidal parameters from (Mikael et al. 2016).

Weld pool width	a	4 mm
Weld pool depth	b	4 mm
Forward weld pool	c_f	4 mm
Rearward weld pool	c_r	4 mm
Forward heat factor	f_f	1.0
Rearward heat factor	f_r	1.0
Exponent constant	n	3
Welding voltage	V	23 V
Welding current	I	80 A
Weld pool energy	Q	5E6 mW
Heat source velocity	v	3 mm/s

3.3 Results of Simulations

The process of modeling the deposition of two beads on a substrate during a Wire Arc Additive Manufacturing (WAAM) process involves multiple steps. Firstly, the geometry of the substrate and welding beads is defined, along with the properties of the materials used. Subsequently, directories are created for each welding and cooling pass to record the temperature, deformation, and history variables specific to each step. The modeling process begins with the simulation of the welding process. The Goldak model is utilized to calculate the temperature distribution generated by the welding bead during deposition. Once the welding pass is completed, the model is used to simulate the cooling process of the welding bead and substrate. Thermal equations and material thermal properties are employed to calculate the temperature distribution during this stage.

Temperature, deformation, and history variables are transferred from one step to another throughout the process, considering the cumulative effects of preceding stages on subsequent ones. For instance, the resulting temperature from welding influences subsequent cooling. At the end of the process, after the deposition of both beads and completion of cooling, the model can be used to calculate the springback on the final structure. This involves evaluating residual deformations and stresses within the structure, accounting for the elastic properties of the materials.

The temperature in the region beneath the electric arc increases rapidly, resulting in a localized melting zone. Figure 9 displays the distributions of vertical displacement, effective plastic strain, and Von Mises stress at the end of the first layer welding. A maximum vertical displacement is observed in the central area of the substrate plate. The plasticized zone is located at the bottom of the first layer and a distortion at the edges. Regions of tension and compression are clearly identified.

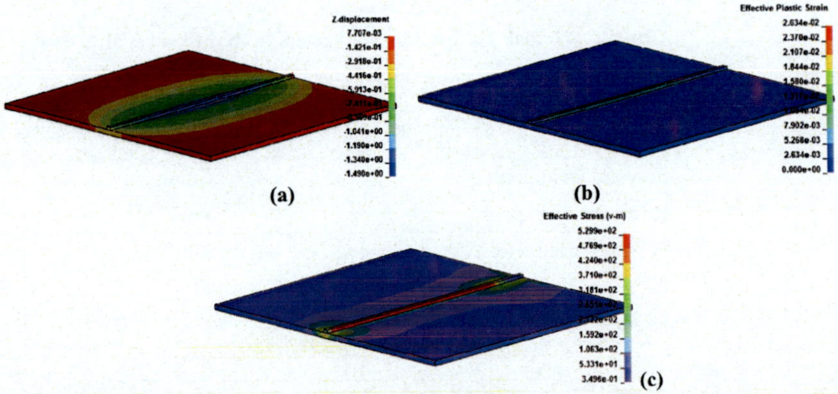

Fig. 9. Distribution after weld pass1 of: (a) vertical displacement, (b) effective plastic strain and (c) Von-Mises stress.

During the cooling of the first layer, heat is transferred through conduction from the hotter regions to the rest of the specimen and dissipated through convection. The thermal contraction of the material during cooling leads to a reduction in effective plastic strain, as shown in Fig. 10.

The complete welding process was simulated by incorporating the welding and cooling of the second layer. Similar results to those of the first layer were observed. The simulation concludes with the calculation of the springback generated upon the release

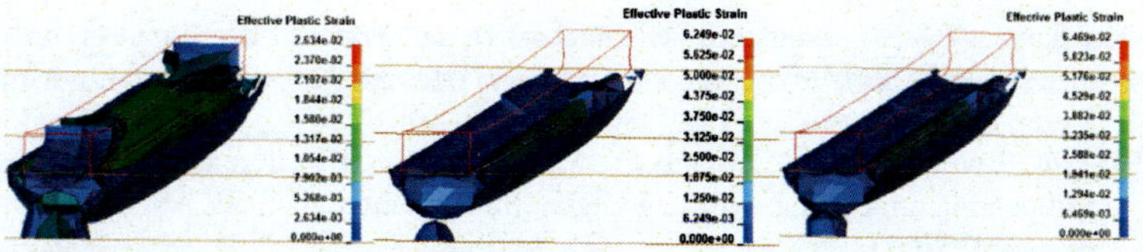

(t=00 sec) (t=1.753 sec) (t=5 sec)

Fig. 10. Effective plastic strain distribution during cooling simulation.

of the substrate plate. The figure depicts the stress field distribution after springback. Figure 11 displays the distributions of vertical displacement and Von Mises stress after springback.

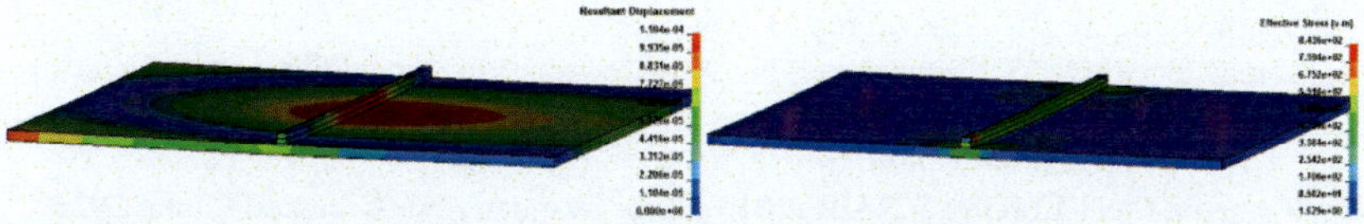

(a) (b)

Fig. 11. Distribution after springback: (a) vertical displacement, (b) Von-Mises stress.

4 Conclusion

The WAAM process is a complex procedure that involves multiple interactions, such as thermal, solid mechanics, and metallurgical effects, among others. Through an extensive experimental study, we have been able to determine the optimal parameters to achieve high-quality welds when employing the WAAM process for the fabrication of complex parts. In addition to the experimental investigation, a numerical model has been developed, taking into account the actual process parameters, including fusion, cooling, and elastic recovery. This numerical model provides a deeper understanding and analysis of the influence of the process on the generation of residual mechanical states in the welded components. To further advance our understanding, several steps can be considered. Firstly, it would be valuable to explore more complex geometries, which would broaden the applicability of the model and provide better insights into the challenges associated with varying shapes. Additionally, conducting an experimental campaign that incorporates in-situ measurements during the process, as well as post-process measurements, would be beneficial in validating and enriching the numerical model.

References

Gurmesa, F.D., Lemu, H.G.: Literature review on thermomechanical modelling and analysis of residual stress effects in wire arc additive manufacturing. Metals **13**(3), 526 (2023). https://doi.org/10.3390/met13030526

Gowthaman, P.S., Jeyakumar, S., Sarathchandra, D.: Effect of heat input on microstructure and mechanical properties of 316L stainless steel fabricated by wire arc additive manufacturing. J. Mater. Eng. Perform., 1–11 (2023). https://doi.org/10.1007/s11665-023-08312-7

Israr, R., Buhl, J., Elze, L., Bambach, M.: Simulation of different path strategies for wire-arc additive manufacturing with Lagrangian finite element methods 15. LS-DYNA Forum 2018, Bamberg (2018)

Kumar, V., Roy, B.K., Mandal, A.: Thermal modeling of wire arc additive manufacturing process using COMSOL multiphysics. In: Dikshit, M.K., Soni, A., Davim, J.P. (eds.) Advances in Manufacturing Engineering: Select Proceedings of ICFAMMT 2022, pp. 223–232. Springer, Singapore (2022). https://doi.org/10.1007/978-981-19-4208-2_16

Mikael, S.M., Jernberg, A., Klöppel, T.: Recent developments for welding simulations in LS-DYNA® and LS-PREPOST®. In: 14th-International-LS-DYNA-Conference, Detroit, USA, 12–14 June 2016 (2016)

Shah, A., Aliyev, R., Zeidler, H., Krinke, S.: A review of the recent developments and challenges in wire arc additive manufacturing (WAAM) process. J. Manuf. Mater. Process. **7**(3), 97 (2023). https://doi.org/10.3390/jmmp7030097

Tomar, B., Shiva, S., Tameshwer, N.: A review on wire arc additive manufacturing: processing parameters, defects, quality improvement and recent advances. Mater. Today Commun. **31**, 103739 (2022). https://doi.org/10.1016/j.mtcomm.2022.103739

Xometry, Steel 1.0038 / S235JR (2015). https://xometry.eu. Accessed 8 June 2023

Zhao, Y., Chen, Y., Wang, Z., Ye, J., Zhao, W.: Mechanical properties, microstructural characteristics and heat treatment effects of WAAM stainless-steel plate material. J. Build. Eng. **75**, 106988 (2023). https://doi.org/10.1016/j.jobe.2023.106988

Parameters Effect Study on Bead Geometry Deposited by CMT Technology Based Wire Additive Manufacturing (WAAM) Process

Oussama Trad[1(✉)], Ated Ben Khalifa[2], Farhat Zemzemi[1], Hédi Hamdi[3], and Borhen Louhichi[1]

[1] LMS, ENISO, University of Sousse, Sousse, Tunisia
oussamatrad01@gmail.com

[2] LGM, ENIM, University of Monastir, Avenue Ibn-Eljazzar, 5019 Monastir, Tunisia

[3] Univ Lyon, Ecole Centrale de Lyon, CNRS, ENTPE, LTDS, UMR5513, ENISE, 42023 Saint Etienne, France

Abstract. Wire-Arc Additive Manufacturing has gained attention due to its ability to build high integrity metal parts. Nevertheless, some challenges persist in this technology, mostly in obtaining optimal deposition parameters that allow for obtaining parts with high geometry quality. The purpose of this study was to investigate the effect of process parameters on the quality of the welded bead made by WAAM deposition strategy based on CMT technology, for samples created using AISI 316L stainless steel welding wire, build up on an AISI 304L stainless steel substrate. A Multiple-Level Factorial Design was carried out, followed by an analysis of variances (ANOVA) to assess the effects of travel speed (TS), wire feed rate (WFR) as well as gas flow rate (GFR) on the beads geometry. The obtained results are promising to be used in future research, such as wall deposition strategy.

Keywords: WAAM process · cold metal transfer (CMT) · ANOVA · Design of experiments (DOE)

1 Introduction

In recent years, there has been a growing interest in using additive manufacturing (AM) for metals to enhance productivity in manufacturing processes across various industries, such as aerospace, shipbuilding, and automotive sectors (Shahrubudin, Lee, and Ramlan 2019). Compared to traditional casting or forging methods, AM offers greater manufacturing productivity and flexibility. One of the most processes used in metal additive manufacturing is wire arc additive manufacturing (WAAM), it involves metal deposition layer by layer coming from a metal wire melted through an electric arc heat source (Derekar 2018). Unlike powder-based additive manufacturing processes that require a laser or electron beam, the WAAM process can utilize a conventional welding machine as the heat source, resulting in lower initial capital costs. The WAAM process is ideal for manufacturing large components with expensive materials, as it can achieve a high

T. Mabrouki et al. (Eds.): AIAM 2023, LNME, pp. 201–211, 2024.
https://doi.org/10.1007/978-3-031-47784-3_24

deposition rate at a lower cost than powder-based processes due to the relatively low cost of wire-type raw materials (about 90% cheaper than powder-type raw materials). Nevertheless, it's worth noting that components produced by WAAM deposition strategy cannot be used in their deposited state due to the presence of overlap deposited areas. Additionally, the WAAM process typically involves post-processing, such as machining or polishing, to meet the final component's requirements (Le et al. 2021). As such, minimizing post-processing requirements can be advantageous to increase productivity, minimizing post-processing requirements is crucial. The buy-to-fly (BTF) ratio can be used to assess the abundance of the WAAM process. This ratio considers the amount of raw material required to produce a component, including the material removed during post-processing. Wire-feed AM processes, such as WAAM, can make parts faster than other metallic additive manufacturing techniques like powder-based systems. But, there is a problem with the quality of the finished product - it may have reduced detail resolution and surface finish, which means that additional machining is required to get the final dimensions and surface quality right. Even with the extra step of machining, WAAM is still more cost-effective than entirely subtractive machining because there is less material waste. The process of making parts using wire-feed AM generates residual stresses and deformation, which can be challenging to control due to the high energy input, deposition rate, and temperature gradient. Therefore, researchers are working hard to understand and control these issues when creating parts. Many process parameters affect directly or indirectly the material performance properties in the WAAM process. Such factors comprise generating power, welding speed, wire feed rate, stand-off distance, gas flow rate, welding trajectory, deposition methods and so forth (Sarathchandra, Davidson, and Visvanathan 2020). This paper models the WAAM process through design of experiments based on a multilevel factorial design. The outputs data, which represent the bead geometry, are evaluated to find the influence of three main operating parameters: wire feed rate (WFR), travel speed (TS) and gas flow rate (GFR).

2 Materials and Methods

The Experimental tests were conducted with a CMT device coupled with a 3-axis CNC machine, while the welding torch was tied to the spindle in a vertical position for the metal deposition as shown in Fig. 1.

A TPS 320i station served to carry out the experimental tests as a source of deposition heat, via a cold metal transfer (CMT) modulus which permits the metal deposition with a low heat input compared to other arc heat units (Dinovitzer et al. 2019). A mixture of argon (Ar) and CO2 was used as a shielding gas. A commercially available SS316L filler wire with a 1 mm diameter was used for deposition process. A 120 × 120 × 6 mm stainless steel plate was used as the substrate for deposition. Table 1, shows the elemental composition of the filler wire.

A synergic mode is used to synchronize and control the current and voltage simultaneously according to the wire feed rate (WFR). The WFR remains proportional to current and voltage, an increase in the WFR indicates an increase in current and voltage values resulting a height generation in the heat input. In this study, four level of the WFR from 3 to 4.5 m/min is used generating a corresponding current and voltage

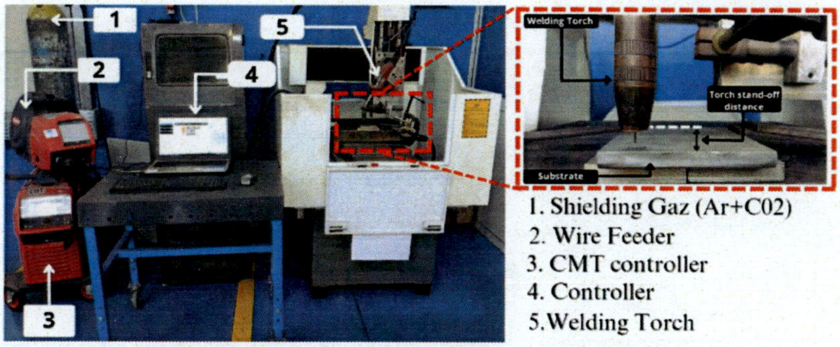

Fig. 1. CMT setup for the WAAM process

Table 1. Elemental composition of AISI 316L welding wire

Element	C	Cu	Cr	Mn	Mo	N	Ni	S	Si	Ti	V	P
Wt. %	0.03	0.75	17.83	2.43	1.62	-	10.58	-	0.62	0.267	0.3	-

that change from 95 A/14.5 V to 117 A/14.8 V respectively. One-bead deposition was carried out on the base plate, based on an experimental design through Multiple-Level Factorial Design (MLFD). For this study research, wire feed rate (WFR), travel speed (TS) and gas flow rate (GFR) were selected as input parameters of the WAAM process. The different process parameters and associated levels are summarized as shown below (Table 2).

Table 2. Different process parameters and their associated levels of the WAAM deposition

Parameters	Level 1	Level 2	Level 3	Level 4
TS (m/min)	0.25	0.5	0.75	1
WFR (m/min)	3	3.5	4	4.5
GFR (L/min)	15	20	-	-

As shown in Fig. 2, all the beads were deposited along the X axis over a length of 50 mm. The stand-off distance (SOD) between the CMT welding torch and the substrate is 12 mm and remains constant for all experiments, with an arc length of 9 mm we obtain 3 mm between the substrate and the wire ends. This was essential to ensure continuous deposition of the material and even good wire feeding into the melting zone. Low stand-off distance values could not ensure an even wire feed due to insufficient space, while high stand-off distance values resulted in wasted wire deposition (Rosli et al. 2020).

A response surface method (RSM) was carried out through multiple-level factorial design for an optimal response via an accurate arrangement of the experimental tests.

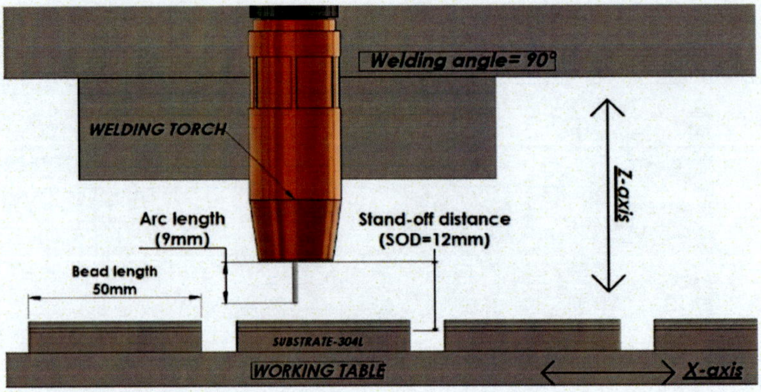

Fig. 2. Schematic representation of the maintained parameters in the experimental tests

Thus, 30 experiments were carried out in the experimental study, Table 3 summarizes the different parameter combinations: wire feed rate (WFR), travel speed (TS), gas flow rate (GFR) and their measured responses, which will be used later for the RSM model.

Several geometric characterization methods can be used to calibrate the weld bead. In order to optimize the welding parameters of the process, in particular travel speed (TS), wire feed rate (WFR) as well as the gas flow rate (GFR), an adjustment of these parameters is necessary to allows controlling the shape of the beads by measuring their geometrical characteristics such as the weld bead height (BH) and the bead width (BW) Fig. 3.

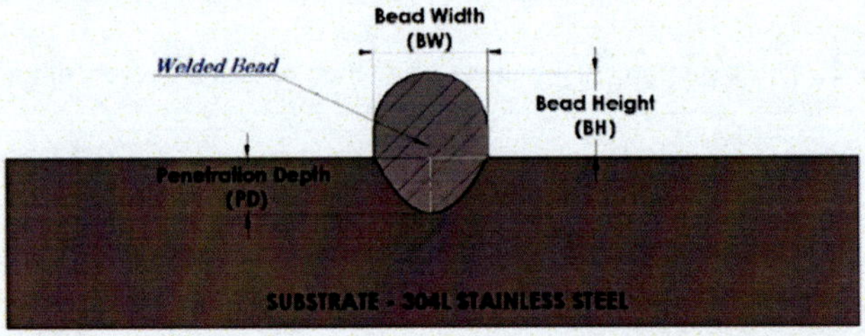

Fig. 3. The measured geometric parameters of welded beads

Several measurements were taken at three separate zones on each sample, averaged to obtain a mean values for evaluation purposes. As mentioned in Fig. 4, the weld bead height (BH) was measured using a welding caliper. For the weld bead width (BW), we use a digital camera to get more accurate measurements than traditional methods.

Table 3. Experimental design and measured outputs

Sample N°	Input parameters			Outputs	
	TS	WFR	GFR	BH	BW
S01	0.75	3.50	20	1.41	1.92
S02	0.50	3.50	15	2.40	2.78
S03	0.25	3.50	20	2.04	4.06
S04	0.75	4.00	20	1.31	2.88
S05	1.00	3.50	15	1.53	2.23
S06	0.50	3.50	20	1.80	2.54
S07	0.50	4.00	20	1.44	2.76
S08	1.00	4.00	15	1.37	2.86
S09	1.00	3.50	20	1.83	1.55
S10	0.50	3.00	15	1.97	1.85
S11	0.75	3.50	15	1.82	3.81
S12	0.50	4.50	15	1.80	2.23
S13	1.00	3.00	20	1.03	2.02
S14	1.00	4.50	15	1.37	3.03
S15	0.75	4.50	20	1.32	3.47
S16	0.50	4.00	15	1.93	3.12
S17	0.25	4.50	15	2.43	5.11
S18	1.00	3.00	15	1.06	2.13
S19	1.00	4.50	20	1.73	2.91
S20	0.25	4.00	20	2.47	4.21
S21	0.50	3.00	20	1.46	2.11
S22	0.25	3.00	20	2.23	2.86
S23	0.75	4.00	15	1.86	2.56
S24	0.25	4.00	15	2.41	4.11
S25	0.25	4.50	20	2.27	5.29
S26	0.75	3.00	15	1.37	2.01
S27	0.75	3.00	20	1.27	2.72
S28	1.00	4.00	20	1.27	2.45
S29	0.25	3.00	15	2.23	2.87
S30	0.50	4.50	20	1.63	4.05
S31	0.25	3.50	15	2.11	4.11
S32	0.75	4.50	15	2.17	3.47

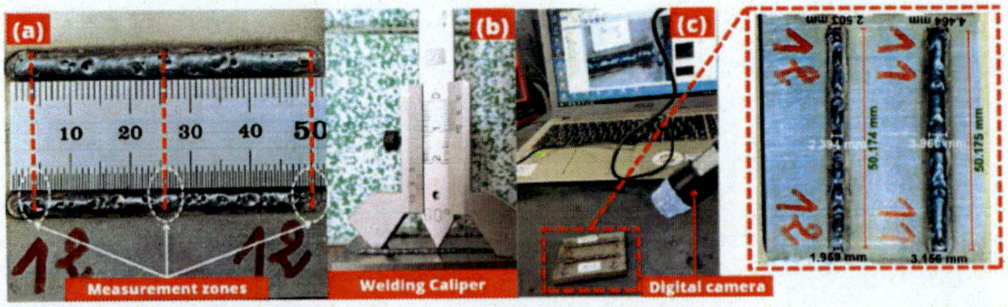

Fig. 4. (a) measurement zones, (b) measurement of the welded bead's height, (c) weld bead width

3 Results and Discussions

In this section, we will explore the results based on the quality of the welded beads. All samples in Fig. 5 and 6 are shown in a perpendicular view to the top surface. Six samples (S05, S10, S13, S18, S26, and S27) formed a discontinuous weld bead. For sample S10 and S13, the droplets remain separate unlike in samples S05, S18, S26, and S27 the droplets are grouped in pairs. On the other hand, some beads have a good weld shape (S01, S07, S09, S24, S25, S29), especially sample S09 which has the best geometry, with good homogenization between the height and width of the bead (BH = 2.27, BW = 5.29) even at the level of the material continuity. In all cases, only S25 bead can be considered acceptable for AM applications due to its good quality.

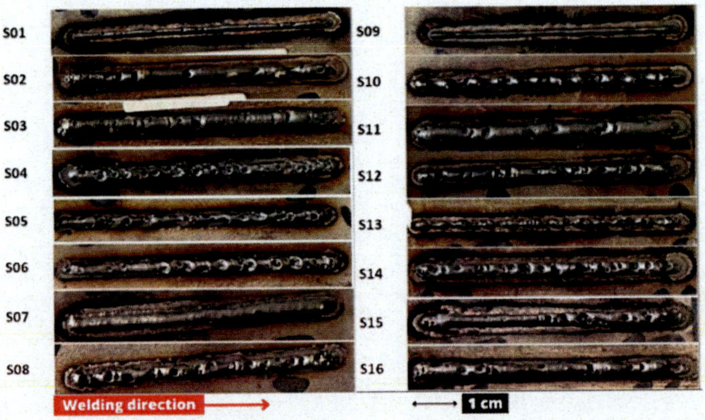

Fig. 5. Perpendicular view of the welded beads S01-S16

3.1 ANOVA Analysis

This section investigates the effect of the wire feed rate (WFR), travel speed (TS) and gas flow rate (GFR) on the beads geometry. Through ANOVA analysis in Table 4, we can find the significant factors which they have the main effect on the response. A P-values

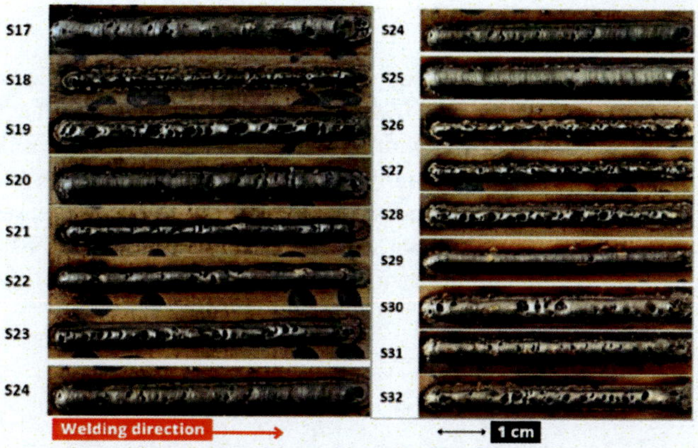

Fig. 6. Perpendicular view of the welded beads S17-S32

between 0 and 0.05 implies that model factors are significant. In this case, travel speed (TS) and wire feed rate (WFR) are significant model terms on the weld bead width (BW). For the bead height (BH), the travel speed (TS) is mainly the most important influencer factor followed by gas flow rate (GFR) in a second order.

Table 4. Analysis of variance for bead geometry response

Bead Width (BW)					
Source	DL	SC adjust	F-value	P-value	
Model	7	20.8919	11.56	0.000	
Linear	7	20.8919	11.56	0.000	
WFR (m/min)	3	7.7865	10.05	0.000	Significant
TS (m/min)	3	13.1026	16.91	0.000	Significant
GFR (L/min)	1	0.0028	0.01	0.918	Non-significant
Bead Height (BH)					
Source	DL	SC adjust	F-value	P-value	
Model	7	4.1458	9.64	0.000	
Linear	7	4.1458	9.64	0.000	
WFR(m/min)	3	0.4048	2.20	0.114	Non-significant
TS (m/min)	3	3.4088	18.50	0.000	Significant
GFR (L/min)	1	0.3321	5.41	0.029	Significant

3.2 Graphical Analysis

Statistical analysis was conducted using Minitab20 software. Selected interval plots are shown in Fig. 7 and 8. The first factor observed is the wire feed rate (WFR), we observe that the bead height and bead width increases as the WFR increases. We also state that for a WFR between 3.5 and 4 m/min, the bead height decreases. Thereafter the travel speed (TS) is examined and in this case, the bead width decreases with travel speed whereas the bead height does not vary significantly. Figure 7 illustrate the influence of the process factors on the bead width evolution (BW).

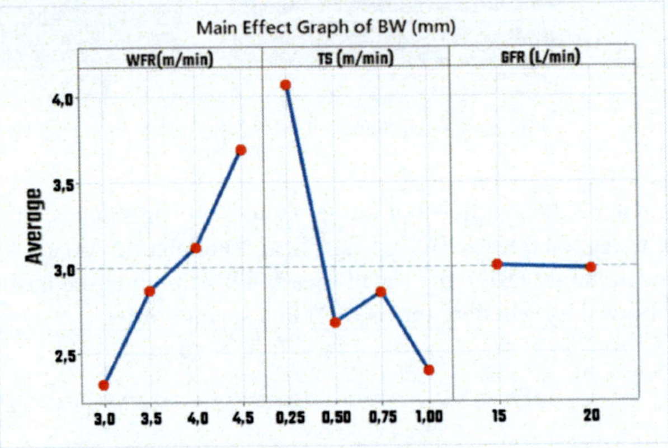

Fig. 7. The effect of the main factors on the bead width (BW)

It is found that the WFR and the TS have an opposite effect on the beads geometry. In Fig. 7, the BW increases when the WFR increases from 3 m/min to 4.5 m/min, in addition we observe a significant fall when the travel speed (TS) change from 0.25 mm/min to 1 mm/min. Travel speed (TS) has the largest impact on the bead width with a contribution

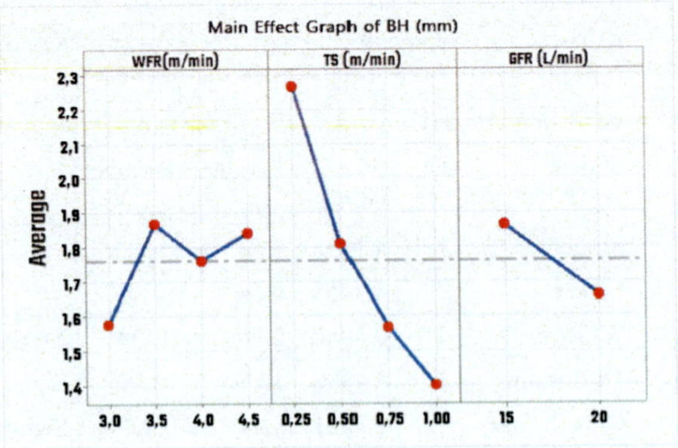

Fig. 8. The effect of the main factors on the bead height (BH)

of 62.62%, coming in second order the WFR (37.22%) and the GFR (0. 03%). In the case of the bead height as represented in Fig. 8, the BH increases when the WFR and the GFR increase from 3 m/min to 3.5 m/min and from 15 L/min to 20 L/min respectively. In contrast, the BH decreases when the TS increase from 0.25 m/min to 1 m/min. The travel speed has the major impact on the bead height accounting for 70.85%. Then the wire feed rate (20.72%) and the gas flow rate (8.42%).

In Figs. 9, 10 and 11, we can observe how the process settings affect the bead geometry, particularly in relation with the bead dimensions (BH, BW). It appears that as the bead width increases with an increase in WFR for all values of the travel speed and the gas flow rate. The beads width consistently falls as the travel speed increase whatever are the values of the wire feed and the gas flow rate. For all values of the travel speed the width only increases when the wire feed rate reaches a value between 4 m/min to 4.5 m/min. The bead height improves when the WFR increases for all values of GFR, however the bead height falls when the travel speed increases for all WFR and GFR values.

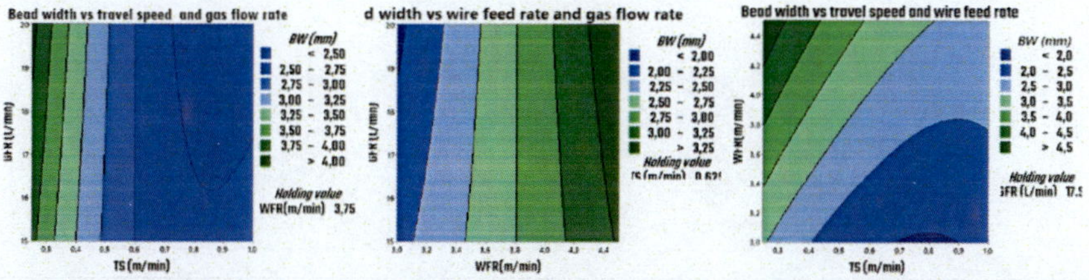

Fig. 9. Contour plots of Bead width (BW) output vs Travel speed (TS), Wire feed rate (WFR) and gas flow rate (GFR)

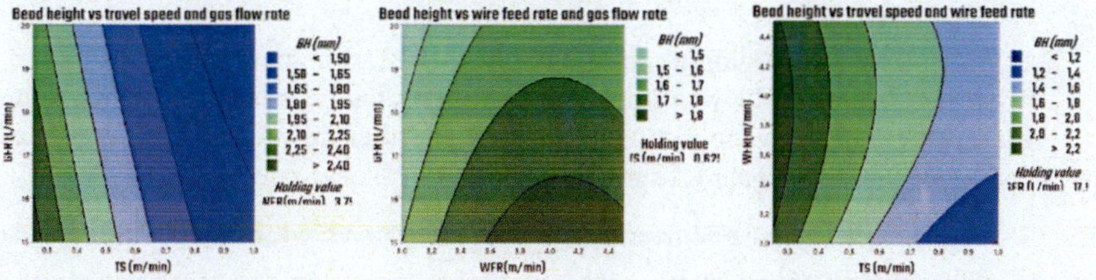

Fig. 10. Contour plots of Bead Height (BH) output vs Travel speed (TS), Wire feed rate (WFR) and gas flow rate (GFR)

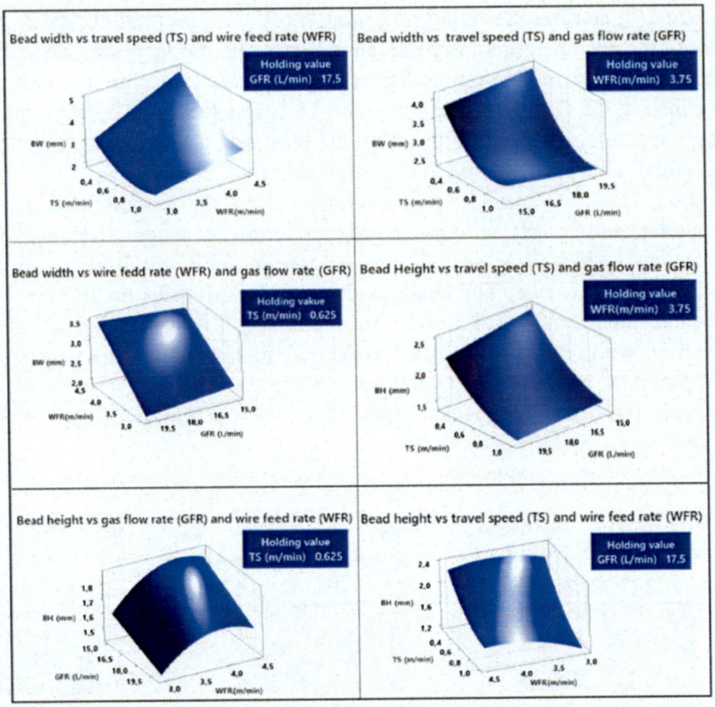

Fig. 11. Response surface of BH and BW outputs vs Travel speed (TS), Wire feed rate (WFR), and gas flow rate (GFR)

4 Conclusion

In the current investigation, AISI 316L single bead manufacturing using the WAAM process based on CMT was studied. An ANOVA analysis was performed to investigate the interactions between individual factors and their response, with significant results found for the investigated responses as follows:

- Bead height increased linearly with wire feed speed, while bead width decreased with increasing travel speed.
- Travel speed (TS) and wire feed have an oppositely related impact on the weld bead dimensions.
- The gas flow rate (GFR) had no significant effect on bead width.
- Travel speed and wire feed had the biggest influential effect on material discontinuity.

Acknowledgements. The authors extend their thanks the National Engineering School of Monastir for supporting the experimental tests. The authors are also grateful to the reviewers Of AIAM'2023 for their helpful comments and recommendations, as these helped to improve this paper.

References

Derekar, K.S.: A review of wire arc additive manufacturing and advances in wire arc additive manufacturing of aluminium. Mater. Sci. Technol. **34**(8), 895–916 (2018). https://doi.org/10.1080/02670836.2018.1455012

Dinovitzer, M., Chen, X., Laliberte, J., Huang, X., Frei, H.: Effect of wire and arc additive manufacturing (WAAM) process parameters on bead geometry and microstructure. Addit. Manuf. **26**, 138–146 (2019). https://doi.org/10.1016/j.addma.2018.12.013

Le, V.T., Mai, D.S., Doan, T.K., Paris, H.: Wire and arc additive manufacturing of 308l stainless steel components: optimization of processing parameters and material properties. Eng. Sci. Technol. Int. J. **24**(4), 1015–1026 (2021). https://doi.org/10.1016/j.jestch.2021.01.009

Rosli, N.A., Alkahari, M.R., Ramli, F.R., Sudin, M.N., Maidin, S.: Influence of process parameters in wire and arc additive manufacturing (WAAM) process. J. Mech. Eng. **17**(2), 69–78 (2020). https://doi.org/10.24191/jmeche.v17i2.15301

Sarathchandra, D.T., Davidson, M.J., Visvanathan, G.: Parameters effect on SS304 beads deposited by wire arc additive manufacturing. Mater. Manuf. Processes **35**(7), 852–858 (2020). https://doi.org/10.1080/10426914.2020.1743852

Shahrubudin, N., Lee, T.C., Ramlan, R.: An overview on 3D printing technology: technological, materials, and applications. Procedia Manuf. **35**, 1286–1296 (2019). https://doi.org/10.1016/j.promfg.2019.06.089

Use of the RDPP-SF Method for Robust Design of Dynamic AM-Fused Deposition Modeling Process

Marwan Amdouni, Atef Dhokkar, Ali Trabelsi(✉), and Mohamed-Ali Rezgui

Laboratoire de Mécanique, Productique et Energétique (LMPE), Ecole Nationale Supérieure d'Ingénieurs de Tunis (ENSIT), 5 Av Taha Hussein, 1008 Montfleury Tunis, Tunisia
ali.trabelsi@rnu.ensit.tn, mohamedali.rezgui@esstt.rnu.tn

Abstract. In manufacturing processes, dynamic systems are extensively used, and finding out the optimal setting of the signal and the operating parameters, which yields the least sensitive design solution about the process responses is a crucial issue. The article proposes a stochastic method (RDPP-SF), which can account for uncertainty in dynamic systems. The method determines the optimal design solution while providing estimates of the natural part-to-part and noise variation components. The virtue of the proposed method is finally demonstrated using an AM-FDM process to monitor the variability in the X, Y, and Z directions of part geometry about a hypothetic CAD model.

Keywords: Dynamic robust design · Stochastic frontier · Uncertainty · Additive manufacture

1 Introduction

Multifunctional engineering teams aim at optimizing superior concepts against uncertainties, which may arise from environmental, manufacturing, batch to batch, and deterioration sources. This is carried out in the parameter design phase to desensitize the process/product against noise factors and their interaction effect with the operating parameters. The Taguchi (Taguchi 1991) method-a pioneering research work in static robust design that has been largely accepted by the engineering and the research communities, can only manage single objective systems, and it does not guarantee the same optimum at each stage unless the mean and the variance of the response are uncorrelated (León et al. 1987). Furthermore, the research works which are developed to extend the employability of the Taguchi method to static and dynamic multi-objective systems proved complex to conceptualize (Maghsoodloo and Chang 2001). This difficulty is partly alleviated by many methods such as the principal component analysis techniques (Su and Tong 1997), the fuzzy logic modeling technique (Huynh et al. 2017), the meta-heuristic methods (Canessa et al. 2014), and the frontier function methods (Al-Refaie et al. 2017). The static RDPP-SF method (Trabelsi and Rezgui 2020; Rezgui and Trabelsi 2020) uses the stochastic frontier model to address the robustness problem. The stochastic frontier paradigm-also known as production function is a parametric technique, which defines the maximum output(s) attainable from a given vector of inputs.

T. Mabrouki et al. (Eds.): AIAM 2023, LNME, pp. 212–218, 2024.
https://doi.org/10.1007/978-3-031-47784-3_25

A dynamic multi-objective system can be stated as in Eq. (1);

$$Y_{js} = f_{js}(X, M_s) + e_{js} \tag{1}$$

where,

- X is the input array and Y_{js} is the output j (j = 1, J) at the signal level Ms = s (s = 1, S);
- The f_{js} is the transfer function (X *vs* Y_{js}) when the signal factor M_s is set at the sth level;
- The e_{js} term represents the uncertainty regarding the Y_{js} response;

The paper is organized as follows; Sect. 2 describes the RDPP-SF method for dynamic multi-objective problems. Section 3 discusses the employability of the proposed method for an AM-FDM application. The computer program, FRONTIER 4.1® is used to estimate the Maximum Likelihood for the parameters of the stochastic frontier model and performs statistical hypotheses about the functional form of the stochastic frontier model.

2 The Dynamic RDPP-SF Method (DRDPP-SF)

The static RDPP-SF method tackles static robust design of products and processes when, among other situations, partial randomization, lack of repeatability/reproducibility in gauging processes, and/or hard-to-change factor are encountered. The RDPP-SF method borrows from the econometric paradigm for the output-oriented production process (Aigner et al. 1977) and it can be stated as in Eq. (2).

$$Y_i = x_i\beta + (v_i - u_i) \qquad i = 1, N \tag{2}$$

where,

- Y_i is the naturally logged output array. The Log transformation assures convexity and stability of the variance,
- xi = [1 ln(x1i) ln(x2i) … ln(xki)2] are the logarithms of the individual and the crossed terms for the ith input, i.e., DMU, i = 1, N,
- β = [β0 β1 … βk]' is a (k + 1) column vector of the unknown parameters for the surrogate model to be estimated by the Maximum Likelihood estimator.
- Vi ~ N(0, σ_v^2) are independently and identically distributed random variables, which represent part-to-part variation, measurement errors, and the combined effect of unspecified input variables,
- Ui ~ N(0, σ_u^2) are independently and identically distributed random variables, which represent the technical inefficiency (TE) for the DMU, i. It reflects the ability of a DMU to obtain maximal output from a given set of inputs. The half-normal distribution for the Ui effects has been most frequently assumed in empirical applications and the best predictor was E(ui|vi − ui).

The output-orientated technical efficiency concept is adopted in this work and answers: "by how much can output quantities be proportionally expanded while keeping

Table 1. Mapping between the econometric and the RDPP-SF method for static systems

Econometric	RDPP-SF method
DMUs	Design plan (DoE)
DMU(i)	Run, i, in the DoE
Panel/cross data	Replicated/non-replicated data
vi: random variation	Natural variation
ui: technical inefficiency	Special cause variation
Metric (RDPP-SF): hypothesis test (95% CI)	
H_0: $\gamma = 0$ vs H_1: $\gamma > 0$ where $\gamma = \frac{\sigma_u^2}{\sigma_v^2+\sigma_u^2}$	
H_0 is rejected if the Likelihood ratio (LR) exceeds $\chi_1^2(2\alpha = 10\%)$	

on the same input quantities". Accordingly, the mapping between the stochastic frontier model and the static RDPP-SF method is shown in Table 1.

The procedural scheme of the dynamic RDPP-SF method for multiple-objective dynamic systems is reviewed as follows.

Step 1 (data preparation): Each performance characteristic (Yi) is transformed and scaled upon the optimization objective, i.e., nominalization, maximization, or minimization. Because the stochastic function originally searches for the maximum output, which can be attained for a combination of the inputs (xi); the outputs of the types, smaller the best (STB) and nominal the best (NTB) should be transformed and scaled beforehand. This is performed using Eqs. (3) and (4).

NTB output is transformed to:

$$Y_i = exp\left[-abs(y_i - y_T)\right] \tag{3}$$

STB output is transformed to:

$$Y_i = \frac{1}{y_i} \tag{4}$$

Step 2 (data analysis): The transfer function for the stochastic frontier model is chosen as a Translog to account for both the interaction and the individual effects of the input factors. The following hypotheses are tested for each output (Y_i) at a 95% confidence level;

a. H_0: $\beta_k = 0$ vs H1: $\beta_k \neq 0$. The test investigates the statistical significance of the main and the interaction effects of the input factors. This is an alternative way to test the fitness of a CD vs a Translog regression model.
b. H_0: $\gamma = 0$ vs H1: $\gamma > 0$. The test investigates the statistical significance of the non-natural variation for each output (Yi). Another way to check the confounding of the stochastic frontier and the neoclassic response surface model (RSM).

c. H_0: U_i ~ half normal vs H_1: U_i ~ truncated normal distribution. The test determines the distributional form of the non-natural variation occurring in the system/process.

Step3 (constitute the uncertainty arrays for the Y_j): The uncertainty array is constituted;

a. If $\gamma \geq 95\%$ then the $v_i \approx 0$. The non-natural variation prevails and the uncertainty array (e_i) is composed of the u_i(s) values. In the FRONTIER4.1® program, the ui(s) terms for each run are obtained as the individual inefficiency (exp(-u)).
b. If $\gamma \leq 5\%$ then the $u_i \approx 0$. The bulk of variation is then due to random unit-to-unit variation. The uncertainty array (e_i) is composed of the v_i(s) estimates at each run in the designed experiment. In this situation, the average line model (response surface model) is confounding with the SF model.
c. If $5\% \leq \gamma \leq 95\%$, both the special cause and the unit-to-unit variations are accountable. The uncertainty array (e_i) is composed of the ($v_i - u_i$) values for each run.

Step 4 (determination of the robust design solution): The uncertainty array for each output, Y_{js}, is constituted based on the γ-value estimate and the inefficiency score of each run. The global uncertainty array is obtained by totting the individual uncertainty scores for the outputs Y_{js} over the S^J factorial combination levels of the signal factor throughout the E executions (runs) of the designed experiment. The combination levels of the input factors (x_{ie}, i = 1, I and e = 1, E) and the signal levels giving the lowest score would correspond to the robust design solution.

3 The Dynamic RDPP-SF Method: A Case Study

The case study is retrieved from the work of Huynh et al. (2017). The AM-FDM process uses the plastic flow of PLA fibers, which have a melting temperature of 180–230 °C and a density of 1.3 kg/cm^3. Two extruders are used. The process uses four operating parameters at three levels each, i.e., x_1: layer thickness to be filled (0.1–0.2–0.3) mm, x_2: print speed (70–80–90) mm/s, x_3: fill density (20–30–40)%, and x_4: nbr. of shells (1–2–3). The signal factor is the building style (1: linear, 2: Hexagonal, 3: diamond). The process outputs are the accuracy of the manufactured specimens on each axis, i.e., $Y_1 = |x_i - x_{i\_CAD})|$. $Y_2 = |y_i - y_{i\_CAD})|$, and $Y_3 = |z_i - z_{i\_CAD})|$. The design strategy is a non-replicated L9 plan as shown on the left side of Table 2. Because the rationale behind the stochastic frontier function drew on the maximization of the production level, the outputs Y_1(DSB), Y_2(DSB), and Y_3(DSB) are transformed as the inverse function. Moreover, to make the example easier for the reader the transfer function is a Cobb-Douglas linear function in x_1, x_2, x_3, and x_4, meaning that we ignore the interaction effects of the input parameters.

Table 2. Dataset of the modified L9 layout with transformed outputs

Run (DMUi)	Factors				$Y_1 = \lvert x_i - x_{i\_CAD} \rvert$			$Y_2 = \lvert y_i - y_{i\_CAD} \rvert$			$Y_3 = \lvert z_i - z_{i\_CAD} \rvert$		
	x_1	x_2	x_3	x_4	S_1	S_2	S_3	S_1	S_2	S_3	S_1	S_2	S_3
1	0.1	70	20	1	71.43	100.0	7.69	4.59	5.38	8.07	13.16	25.00	9.43
2	0.1	70	20	2	6.76	10.87	4.90	3.85	4.51	7.81	7.69	7.81	6.25
3	0.1	70	20	3	7.25	6.02	4.43	5.05	6.33	9.26	15.63	26.32	18.52
4	0.2	90	30	1	3.94	10.64	5.81	11.63	10.20	15.63	38.47	15.63	8.33
5	0.2	90	30	2	3.70	5.68	4.24	10.20	13.51	10.87	10.64	17.24	45.46
6	0.2	90	30	3	5.16	5.32	5.16	7.81	8.77	6.67	14.29	13.89	25.00
7	0.3	80	40	1	4.67	5.88	4.31	5.21	5.26	5.05	19.23	20.83	12.20
8	0.3	80	40	2	4.72	4.85	6.17	6.94	6.49	4.90	10.42	17.86	10.87
9	0.3	80	40	3	8.20	12.50	12.5	3.70	4.00	3.70	13.51	25.00	12.82

4 Discussion and Findings

The uncertainty score estimate, $Abs(u_i - v_i))$, at each run, i, represents the magnitude of the global variation while considering the setting of the operating factors and the signal level. The loglikelihood estimator and the hypotheses tests on the γ-value are performed to evaluate the u_i and v_i scores for each run, i.

As stated earlier, ui and vi represent the part-to-part and the noise variations in log scaling, respectively. Table 3 shows the $Abs(u_i - v_i)$ estimates at different levels of the signal factor.

Afterward, the uncertainty scores in each run, i are totted over the factorial combinations of the signal levels for the outputs Y_1, Y_2, and Y_3, individually.

Table 4 shows the global uncertainty scores at different factorial combinations across the signal levels. Only potential solutions, among the 3^3 combinations, are shown in Table 4. The minimum global uncertainty value is found to be 0.815, meaning that the robust design solution corresponds to run 1 ($x_1 = 0.1$ mm, $x_2 = 70$ mm/s, $x_3 = 20\%$, and $x_4 =$ one shell). The signal levels for the outputs Y_1, Y_2, and Y_3, however, should be set at $S_3 =$ Diamond for Y_1, $S_1 =$ Linear for Y_2, and $S_3 =$ Diamond for Y_3, respectively. Table 4 also shows that other potential design solutions such as run 1 ($S_2S_1S_3$), run 3 ($S_1S_1S_3$), and run 1 ($S_2S_1S_3$) are worth considering for economic trade. In case the same signal level should be adopted for all outputs, equivalently, the least sensitive solution would be run 9 ($S_1S_1S_1$) or run 1 ($S_3S_3S_3$), which have global uncertainty scores of 1.794 and 1.069, respectively (see Table 4).

Table 3. Uncertainty arrays at different signal factor

Run (DMUi)	Y_1			Y_2			Y_3		
	εi = abs(vi − ui)			εi = abs(vi − ui)			εi = abs(vi − ui)		
	S_1: Lin.	S_2: Polyg.	S_3: Diam.	S_1: Lin.	S_2: Polyg.	S_3: Diam.	S_1: Lin.	S_2: Polyg.	S_3: Diam.
1	2.214	0.043	0.001	0.446	2.893	0.700	1.393	12.413	0.368
2	1.780	1.786	1.153	0.224	0.586	0.944	0.929	14.629	1.142
3	1.646	2.097	1.665	0.214	0.055	0.925	0.777	15.217	1.433
4	0.826	1.092	0.654	0.824	0.166	0.990	1.549	15.033	1.755
5	0.642	1.243	1.671	0.854	0.881	1.099	1.336	15.657	2.394
6	0.280	1.030	1.885	0.616	0.998	0.783	0.956	15.721	2.387
7	1.090	1.529	2.337	0.021	0.032	0.340	1.115	15.406	1.707
8	0.988	1.245	2.679	0.090	0.311	0.226	0.984	15.594	1.671
9	0.863	0.020	0.081	0.090	0.583	3.355	0.842	14.646	1.148

Table 4. Global uncertainty scores for factorial combinations across the factor signal levels

Run (DMUi)	Combination of signal levels										
	$S_1S_1S_1$	…	$S_2S_1S_1$	…	$S_2S_1S_2$	$S_2S_1S_3$	$S_2S_2S_2$	…	$S_3S_1S_3$	…	$S_3S_3S_3$
1	4.053	…	1.881	…	12.901	<u>0.856</u>	15.349	…	<u>0.815</u>	…	**1.069**
2	2.932	…	2.938	…	16.638	3.151	17.001	…	2.518	…	3.238
3	2.637	…	3.089	…	17.528	3.744	17.369	…	3.312	…	4.023
4	3.199	…	3.465	…	16.949	3.671	16.291	…	3.234	…	3.399
5	2.832	…	3.433	…	17.754	4.491	17.780	…	4.920	…	5.165
6	1.853	…	2.603	…	17.367	4.034	17.749	…	4.889	…	5.055
7	2.226	…	2.665	…	16.956	3.257	16.966	…	4.065	…	4.384
8	2.062	…	2.319	…	16.929	3.006	17.150	…	4.440	…	4.575
9	**1.794**	…	0.952	…	14.756	1.258	15.250	…	1.319	…	4.584
Min	1.794	…	0.952	…	12.901	0.856	15.250	…	0.815	…	1.069

5 Conclusion

The proposed method has employed the stochastic frontier model for the robust design of a dynamic AM-FDM process. The robust solution is Runs 1 (x_1: layer thickness is set at 0.1 mm, x_2: print speed is set at 70 mm/s, x_3: fill density is set at 20%, and x_4: nbr. of shells is one). Setting of the signal levels is, S_1 = Diamond, S_2 = Linear, and S_3 = Diamond, for the Y_1, Y_2, and Y_3 outputs, respectively. The optimal global uncertainty score is exp(0.815) = 2.259 mm in linear scaling, which is distributed, exp(0.00131) =

1.001 mm in the X direction; exp(0.4455) = 1.561 mm in the Y direction, and exp(0.368) = 1.445 mm in the Z direction. When considering the same signal level for all Y_i, the global uncertainty score is 1.050 (2.858 in linear scaling) and the robust solution corresponds to run 3 (x_1: layer thickness is set at 0.1 mm, x_2: print speed is set at 70 mm/s, x_3: fill density is set at 20%, and x_4: nbr. of shells is three). The signal levels for the outputs Y_1, Y_2, and Y_3, however, should be established at S_1 = Diamond, S_2 = Linear, and S_3 = Diamond, respectively.

Acknowledgments. The authors are grateful for the valuable support of the AIAM2023 scientific committee. We highly appreciate the reviewers' insightful and helpful comments on our manuscript.

References

Taguchi, G.: Systems of Experimental Design, 4th edn. vol. 1, 2. American Supplier Institute, Dearborn, MI (1991)

León, R.V., Anne, C.S., Kacker, R.N.: Performance measures independent of adjustment: an explanation and extension of Taguchi's signal-to-noise ratios. Technometrics **29**(3), 253–265 (1987)

Maghsoodloo, S., Chang, C.L.: Quadratic loss functions and signal-to-noise ratios for a bivariate response. J. Manuf. Syst. **20**(1), 1–12 (2001)

Su, C.T., Tong, L.I.: Multi-response robust design by principal component analysis. Total Qual. Manag. **8**(6), 409–416 (1997)

Huynh, H.N., Nguyen, A.T., Ha, N.L., Thai, T.T.H.: Application of fuzzy taguchi method to improve the dimensional accuracy of fused deposition modeling. In: International Conference on System Science and Engineering (ICSSE), pp. 107–112. IEEE (2017)

Canessa, E., Bielenberg, G., Allende, H.: Robust design in multiobjective systems using Taguchi's parameter design approach and Pareto genetic algorithm. Rev. Fac. Ing.-Univ. Ant. **72**, 73–85 (2014)

Al-Refaie, A., Al-Alaween, W., Diabat, A., et al.: Solving dynamic systems with multi-responses by integrating desirability function and data envelopment analysis. J. Intell. Manuf. **28**, 387–403 (2017)

Trabelsi, A., Rezgui, M.A.: Robust design of processes and products using the mathematics of the stochastic frontier. Int. J. Adv. Manuf. Technol. **106**(7–8), 2829–2841 (2020)

Rezgui, M.A., Trabelsi, A.: Use of the stochastic frontier and the grey relational analysis in robust design of multi-objective problems. Concurrent Eng.-Res. A **28**(2), 110–129 (2020)

Aigner, D., Knox Lovell, C.A., Peter, S.: Formulation and estimation of stochastic frontier production function models. J. Econ. **6**(1), 21–37 (1977)

Effect of Printing Process Parameters on the Tensile Behavior of FDM-Built Parts

Asma Belhadj[1(✉)], Salma Slama[1], Mouhamed Hichem Habouba[2], and Tarek Mabrouki[1]

[1] National School of Engineers of Tunis (ENIT), Applied Mechanics and Engineering Laboratory (LMAI-LR-11-ES19), University of Tunis El Manar (UTM), Tunis, Tunisia
{asma.belhadj,selma.slama,tarek.mabrouki}@enit.utm.tn
[2] AKWEL Society, Mateur, Tunisia
hichem.habouba@akwel-automotive.com

Abstract. The present study deals with the qualification of the effect of polymer printing parameters on tensile mechanical properties of parts manufactured based on the fused filament deposition (FDM) process. Two types of polymers, namely PLA and PETG, were selected for this study. For that the influence of three key FDM parameters which are layer thickness, infill ratio and infill pattern on the mechanical behavior of precited parts were evaluated according a design of experiment. Particular attentions were taken to Young's Modulus (E) and Ultimate Tensile Strength (UTS) evolutions. The results demonstrate that these three process parameters have an important influence on the variation of the tensile properties of FDM-manufactured parts. Furthermore, optimal printing parameters for PLA and PETG were determined resulting in higher mechanical properties than those obtained using parameters indicated in the filament technical datasheets.

Keywords: FDM · PLA · PETG · Experimental design · mechanical properties

1 Introduction

Manufacturing techniques are constantly advancing to meet the evolving demands and needs of various industries. Among these techniques, Fused Deposition Modeling (FDM) has gained popularity as a cost-effective and straightforward additive manufacturing (AM) method. FDM utilizes melted polymeric material in the filament form that is extruded through a heated nozzle to create 3D parts-built line by line. The quality of mechanical properties of the FDM-manufactured parts depends on process parameter's selection (Mohamed et al. 2015, Gordelier et al. 2019).

Although FDM was one of the first processes to be invented, it still holds great interest in the scientific community. In fact, the quality of parts obtained through FDM can vary considerably depending on the choice of many parameters involved. Several studies have been published on this subject since the early 2000s. ABS polymer has historically been the most extensively studied material (Es-Said et al., 2000; Lee et al., 2005; Durgun and Ertan, 2014). More recently, there has been a surge in interest in PLA

T. Mabrouki et al. (Eds.): AIAM 2023, LNME, pp. 219–227, 2024.
https://doi.org/10.1007/978-3-031-47784-3_26

polymer, whereas research on PETG remains relatively limited. Lanzotti et al. 2015 have analyzed FDM process parameter's effect on the evolution of tensile behavior of PLA (Lanzotti et al. 2015). These authors have varied three manufacturing parameters: layer thickness (0.1 mm; 0.12 mm; 0.15 mm; 0.18 mm; 0.2 mm), filling orientation (0°; 18°; 54°; 72°; 90°) and the number of contours (2; 3; 4; 5; 6). They validated the fact that the Young's modulus and the tensile strength are maximized by the optimal combination of the following parameters: layer height 0.2 mm, filament orientation 0°, and number of outer contours equal to 3. Additionally, Rajpurohit et al. also studied the influence of three manufacturing parameters by FDM: the screen angle (0°, 30°, 45°, 60° and 90°), the screen thickness (0.4 mm, 0.5 mm, 0.6 mm and 0.7 mm) and the layer thickness (0.1 mm, 0.15 mm, 0.2 mm, 0.25 mm and 0.3 mm) on the mechanical tensile properties of PLA. The samples are produced in accordance with the ASTM D638 standard, at an extrusion temperature of 210 °C and 100% filling. They found that the highest tensile strength (i.e. 47.36 ± 2.69 MPa) is obtained at a screen angle of 0°, a screen width of 0.6 mm and a layer thickness of 0.1 mm (Rajpurohit et al. 2018). Sirnivasan et al. investigated the effect of the infill ratio on the ultimate tensile strength and the roughness of PETG specimens obtained by FDM keeping other fabrication parameters constant (Sirnivasan et al. 2020). They found that by increasing the fill rate, the tensile strength gradually increases. For 100% infill ratio, they obtained the highest tensile strength value of 32.12 MPa. For 20% infill ratio, they found a tensile strength value of 17.38 MPa. This is mainly due to the presence of vacuum in the part obtained by additive manufacturing. On the other hand, they found that surface roughness increases with decreasing % infill ratio. The 100% ratio presented the best roughness value of 2.87 μm.

In this study, an experimental methodology is presented to investigate the impact of three polymer printing parameters (the layer thickness, the infill pattern and the infill ratio) on the tensile mechanical properties' variations of FDM-produced parts. The aim of this study is to determine FDM process parameters to be used for enhancing the Young's modulus (E) and ultimate tensile strength (UTS) of PLA and PETG polymer parts.

2 Experimental Protocol

2.1 FDM Process and Materials

Developed in 1989 by Scott and Lisa Crump, the Fused Deposition Modeling (FDM) process is based on the layer-by-layer deposition of a liquefied filament material (polymer) using a heated nozzle that moves along x and y. The part is manufactured on a build platform guided in translation along the z axis (Fig. 1). The FDM process involves several printing parameters that can be adjusted to optimize the final part quality. Some of the most common FDM parameters are:

- **Layer height**: corresponds to the vertical thickness of each layer in the printed part.
- **Infill ratio**: corresponds to the material amount inside the part, usually represented as a percentage.
- **Infill pattern**: corresponds to the internal structure pattern of the part.
- **Printing temperature**: corresponds to the temperature at which the thermoplastic material is melted and extruded.

- **Printing speed**: corresponds to the rate at which the printer head traverses during the printing process.

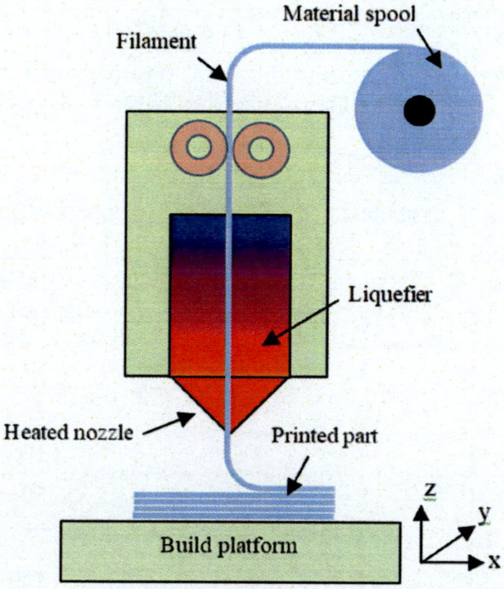

Fig. 1. FDM process

The used materials are two types of polymers: PLA and PETG. The PLA, also known as poly-lactic acid, is an eco-friendly thermoplastic. It is a first natural material used in AM. For this study, sky blue PLA-HI OWA is used. The PETG, or glycol-modified polyester, is a thermoplastic that combines the properties of PET (polyethylene terephthalate) and glycol. The glycol element reduces the brittleness of PET and this minimizes its fragility. PETG is also widely used in the AM market, particularly in FDM, owing to its ease of printing and high resistance. In this study, the opaque white PETG Ultra is used.

2.2 Experimental Design

In this study, experimental design with three factors at three levels is used. The choice of parameters and their levels, presented in Table 1, is based on recent published works (Rajpurohit and Dave 2018, Fekete et al. 2021). This has enabled us to define a Taguchi's experimental plan, and 9 experimental runs were made based on the design matrix exposed in Table 2.

Table 1. Parameters levels

Parameter	Level 1	Level 2	Level 3
Layer height (mm)	0,2	0,3	0,4
Infill ratio (%)	50%	75%	100%
Infill pattern	Grid (G)	Honeycomb (N)	Triangular (T)

Table 2. Matrix of experiments (L9 Taguchi)

	Parameter			Layer height (mm)	Infill ratio (%)	Infill pattern
	A	B	C			
1	1	1	1	0,2	50	Grid
2	1	2	2	0,2	75	Honeycomb
3	1	3	3	0,2	100	Triangular
4	2	1	2	0,3	50	Honeycomb
5	2	2	3	0,3	75	Triangular
6	2	3	1	0,3	100	Grid
7	3	1	3	0,4	50	Triangular
8	3	2	1	0,4	75	Grid
9	3	3	2	0,4	100	Honeycomb

2.3 Samples Manufacturing

Tensile tests were conducted on specimens whose shape and dimensions, shown in Fig. 2, were taken from the ISO 527 standard. The specimens were manufactured using a 3D printer, the Volumic Stream 30 Pro MK2 (Fig. 3), provided by AKWEL. In this work, fixed printing parameters are used for the two polymers, and they are presented in Table 3.

2.4 Tensile Tests

In this study, tensile tests are conducted using a Zwick Roell Z005 tensile machine at the AKWEL company, shown in Fig. 4b. The positioning of the specimens is done by flat jaws tightened by pneumatic controls. The test conditions are defined by the ISO 527-2 standard. The test speed is 1 mm/min and the distance between the jaws is 115 mm. The environmental conditions of the tests were those of the laboratory: an atmosphere containing about 50% humidity and a temperature of about 25°. For each test, 5 specimens are made, according to ISO 527-2: 2012 standard (Fig. 4a).

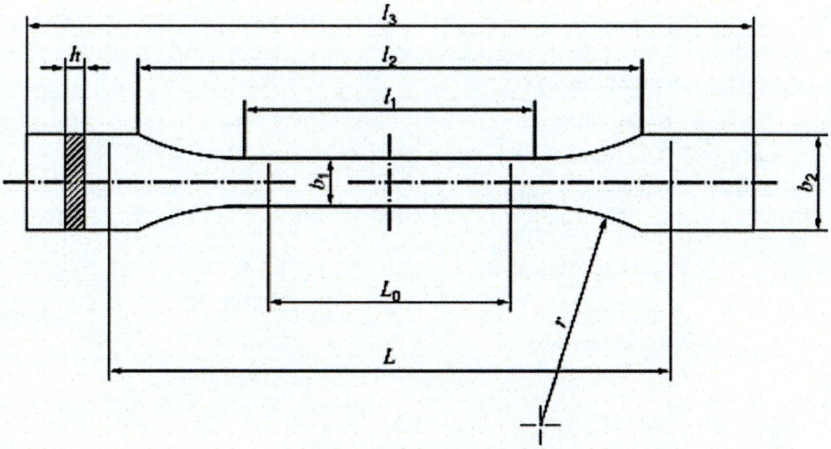

Fig. 2. Samples shape and dimensions (ISO 527-2 – 2012)

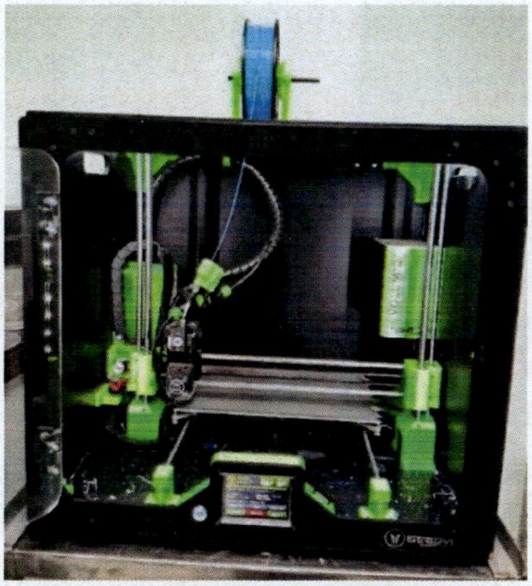

Fig. 3. FDM Machine

3 Results and Discussion

From the obtained tensile curves, the average value of the elastic modulus (E) and UTS are determined for each printing test for both PLA and PETG materials.

The histograms of Fig. 5 present the average elastic modulus values for the 9 tests carried out. In the Fig. 5a, it can be clearly seen that the modulus of elasticity of the PETG increases according the infill ratio. It can also be observed that a infill ratio of 100% with the used patterns considerably improves the modulus of elasticity. Indeed, with 100% infill ratio, the Young's modulus of the specimens exceeds the elastic modulus obtained with the parameters recommended in the filament technical datasheet, which is equal to 1198 ± 84 MPa. The highest Young's modulus (1694.52 MPa) is achieved in test 3 for

Table 3. Fixed printing parameters.

Fixed parameters	Value
Nozzle diameter	0.4 mm
Deposition speed	60 mm/s
Number of perimeters	3
Perimeter pattern	Rectilinear
Build plate temperature	PLA 50 °C
	PETG 60 °C
Nozzle temperature	PLA 200 °C
	PETG 220 °C

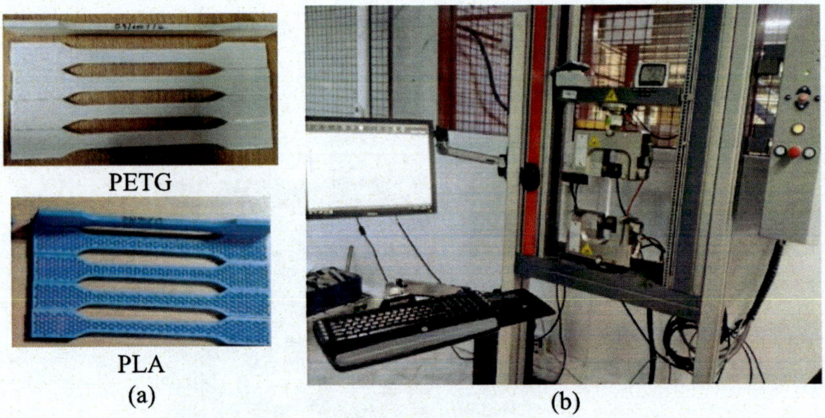

Fig. 4. a) Tensile samples. b) Tensile test machine

a specimen obtained with a 0.2 mm layer thickness, a 100% infill ratio and a triangular patter. In the Fig. 5b, it is worth noting that elastic modulus E of PLA increases with the infill ratio. Elastic modulus with the parameters recommended in the filament technical datasheet is equal to 2491 MPa. Moreover, it can be underlined that 100% infill ratio improves the elastic modulus of the FDM specimens, reaching 3649 MPa in the case of the specimen obtained with a 0.2 mm layer thickness, a 100% ratio and a triangular pattern.

The histogram of Fig. 6 shows the mean values of ultimate tensile strength for each test. For PETG, the tensile strength of printed part increases according the infill ratio. However, this increase is not linear. In fact, the difference between the ultimate strength of 75% and 100% infill ratio is greater than the difference between 50% and 75%. The highest ultimate tensile strength is reached in test 6 (0.3/100%/Grille) with a value of 39.27 MPa (Fig. 6a). Furthermore, it can be seen that the 100% filling rate with the different used patterns exceeds the ultimate strength value given in filament data sheet

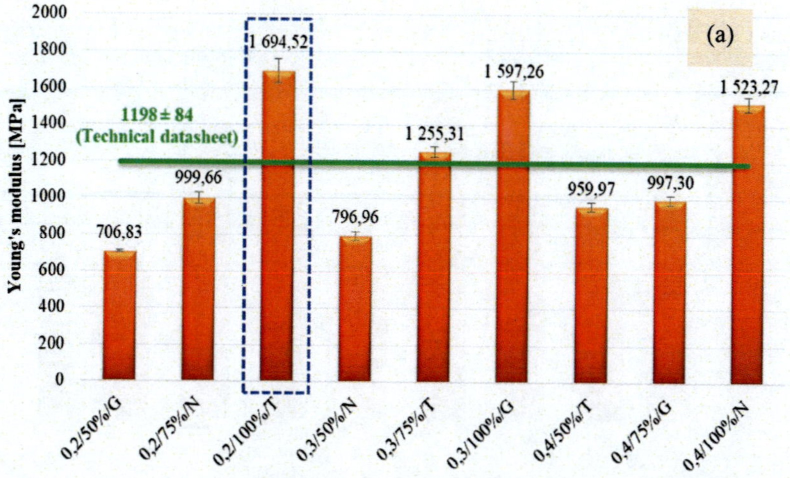

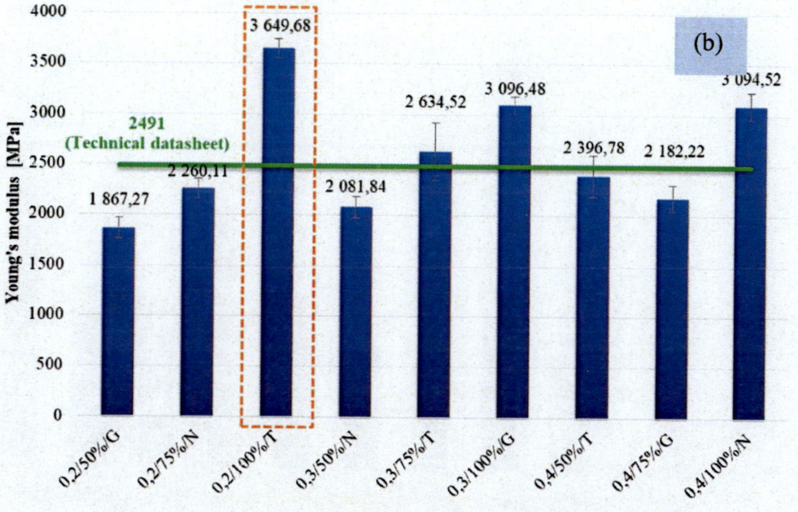

Fig. 5. Average Young's modulus for: (a) PETG and (b) PLA

(26 ± 1 MPa). For printed PLA, the histogram of Fig. 6b shows that the ultimate tensile strength increases as a function of the infill ratio, nonlinearly. The highest maximum stress (38.79 MPa) is obtained for a 0.2 mm layer thickness, a 100% ratio and a triangular pattern.

By analyzing the obtained results, it can be underlined that:

- The layer thickness has less effect on E and UTS of PETG and PLA.
- The elastic modulus E and tensile strength of PETG and PLA materials demonstrate an increasing trend as the infill ratio increases. Notably, the increase in infill ratio has a significant effect on the modulus of elasticity. This can be attributed to the fact that

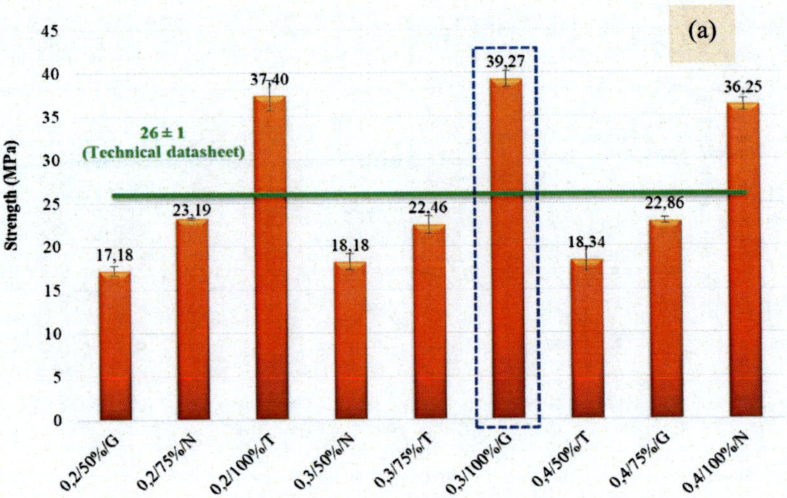

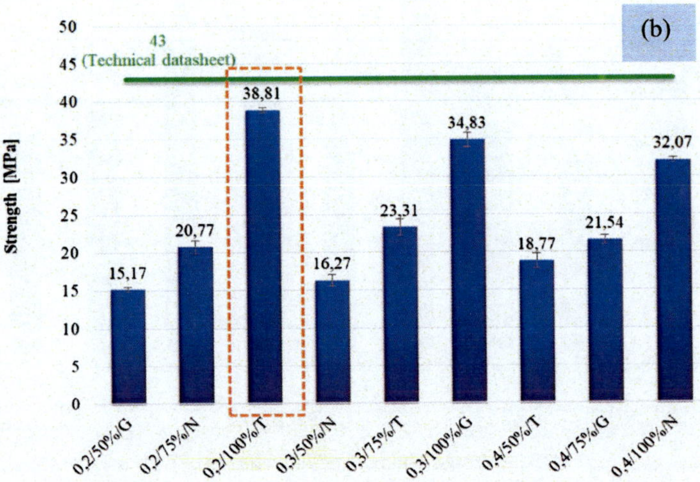

Fig. 6. Average ultimate tensile strength for: (a) PETG and (b) PLA

a higher ratio results in reduced voids within the printed part. As a result, the printed part becomes stiffer, leading to a higher E.

- The triangular infill pattern involves the highest E and UTS for PLA.
- The triangular infill pattern results in the highest E for PETG.
- The infill pattern hasn't impacted the tensile strength of PETG.

4 Conclusion

This study explores the mechanical characteristics of polymer components (PETG and PLA) fabricated using the FDM technique. A design of experiment based on Taguchi plan was adopted to study influences of layer thickness, infill ratio and infill pattern on some tensile proprieties of built FDM-parts. Tensile tests were carried out in concordance with ISO-527 standards to determine the elastic modulus and the UTS of FDM printed parts. The optimal process parameters improving stiffness for PLA and PETG is 0.2 mm as layer height, 100% as infill ratio and triangular as infill pattern. For the strength, the optimal parameters combinations are (0.3/100%/Grid) for PETG and (0.2/100%/Triangular) for PLA.

References

Es-Said, O.S., Foyos, J., Noorani, R., et al.: Effect of layer orientation on mechanical properties of rapid prototyped samples. Mater. Manuf. Process. **15**(1), 107–122 (2000). https://doi.org/10.1080/10426910008912976

Lee, B.H., Abdullah, J., Khan, Z.A.: Optimization of rapid prototyping parameters for production of flexible ABS object. J. Mater. Process. Technol. **169**(1), 54–61 (2005). https://doi.org/10.1016/j.jmatprotec.2005.02.259

ISO En 527-1: Plastics—Determination of tensile properties—Part 1: General principles. International Organization of Standardization: Geneva, Switzerland (2012)

ISO En 527-2: Plastics-Determination of tensile properties-Part 2: Test conditions for molding and extrusion plastics. International Organization for Standardization: Switzerland (2012)

Durgun, I., Ertan, R.: Experimental investigation of FDM process for improvement of mechanical properties and production cost. Rapid Prototyp. J. **20**(3), 228–235 (2014). https://doi.org/10.1108/RPJ-10-2012-0091

Lanzotti, A., Grasso, M., Staiano, G., et al.: The impact of process parameters on mechanical properties of parts fabricated in PLA with an open-source 3-D printer. Rapid Prototyp. J. **21**(5), 604–617 (2015). https://doi.org/10.1108/RPJ-09-2014-0135

Mohamed, O.A., Masood, S.H., Bhowmik, J.L.: Optimization of fused deposition modeling process parameters: a review of current research and future prospects. Adv. Manuf. **3**(1), 42–53 (2015). https://doi.org/10.1007/s40436-014-0097-7

Rajpurohit, S.R., Dave, H.K.: Effect of process parameters on tensile strength of FDM printed PLA part. Rapid Prototyp. J. **24**(8), 1317–1324 (2018). https://doi.org/10.1108/RPJ-06-2017-0134

Gordelier, T.J., Thies, P.R., Turner, L., et al.: Optimizing the FDM additive manufacturing process to achieve maximum tensile strength: a state-of-the-art review. Rapid Prototyp. J. **25**(6), 953–971 (2019). https://doi.org/10.1108/RPJ-07-2018-0183

Srinivasan, R., Ruban, W., Deepanraj, A., et al.: Effect on infill density on mechanical properties of PETG part fabricated by fused deposition modelling. Mater. Today Proc. **27**, 1838–1842 (2020). https://doi.org/10.1016/j.matpr.2020.03.797

Fekete, I., Ronkay, F., Lendva, L.: Highly toughened blends of poly (lactic acid) (PLA) and natural rubber (NR) for FDM-based 3D printing applications: the effect of composition and infill pattern. Polym. TestingTesting **99**, 107–205 (2021). https://doi.org/10.1016/j.polymertesting.2021.107205

Experimental Investigation of SLM Parameters Effects on Roughness of 316L Parts

Oussema Bouguerra[1(✉)], Salma Slama[2], Asma Belhadj[2], and Noureddine Barka[1]

[1] Mathematics, Software and Engineering Department, University of Quebec at Rimouski, Rimouski, Canada
{Oussema.Bouguerra,Noureddine_Barka}@uqar.ca
[2] LMAI, ENIT - UTM, BP 37 Le Belvédère, 1002 Tunis, Tunisia
{selma.slama,asma.belhadj}@enit.utm.tn

Abstract. This work is an experimental parametric study that employs the design of experiments methodology to investigate the influence of printing parameters on the surface quality of 316L stainless steel parts manufactured through the SLM process. To achieve this objective, a full factorial plan consisting of three factors at three levels is used. The factors investigated are laser power (P), scanning speed (V) and hatching distance (HD). This study involves assessing the impact of these parameters on the roughness of SLM printed parts and determining the various combinations that result in a good surface condition.

Keywords: Parametric study · SLM · 316L stainless steel · Roughness

1 Introduction

Selective Laser Melting (SLM) is the most used process in metal additive manufacturing due to its ability to produce high-performance components with complex geometries. This process involves melting and solidifying successive layers of metal powder to create a volumetric part. It is widely employed in industries such as aerospace, automotive, biomedical, and space sectors. However, the quality of the printed parts, as well as their mechanical properties, depends on the chosen printing parameters (Ahmed et al. 2022, Greco et al. 2020).

Miranda et al. 2016 conducted a study on Selective Laser Melting (SLM) and found that increasing the laser power resulted in a gradual reduction of porosity, ultimately leading to improved density of the manufactured parts. Hao et al. 2006 demonstrated that there exists an optimal scanning speed value at which the tensile strength is maximized. Deviating from this optimum value can result in reduced tensile strength. Li et al. 2010 conducted research on the effects of process parameters on porosity and consequently material density. Their study aimed to understand how adjusting various parameters influenced the level of porosity in the printed parts, thereby affecting material density. Greco et al. 2020 stated that a higher layer thickness in additive manufacturing processes leads to a decrease in density and a reduction in microhardness. Their research suggested that layer thickness can have a significant impact on the overall quality and

T. Mabrouki et al. (Eds.): AIAM 2023, LNME, pp. 228–237, 2024.
https://doi.org/10.1007/978-3-031-47784-3_27

mechanical properties of the printed parts. According to Tucho et al. 2018, a smaller hatch spacing ensures that the melting basins overlap and cover each other more effectively. This overlapping of melting basins leads to a reduction in pore formation during the printing process. Their research highlights the importance of optimizing the hatch spacing parameter to minimize porosity and enhance the quality of the printed parts.

This work aims to investigate the effect of the principle SLM parameters on the quality of 316L stainless steel parts. An experimental parametric study was conducted using a complete three-factor design with three levels. The studied parameters are laser power (P), scanning speed (V) and hatching distance (HD). The different factor levels were selected based on previous research and work conducted at the UQAR laboratory. The results of this study enabled us to determine the optimal parameter combinations that yield high-quality surface conditions for printed parts.

2 Principle of the SLM Process

Selective laser melting (SLM) is an additive manufacturing technique developed specifically for 3D printing metal alloys. It involves the fusion of metal powder particles in a process known as total melting to create additive parts. The 3D-printed object is constructed layer by layer, starting from a 3D model. The process begins by depositing a layer of powder on the build platform. Then, a laser beam selectively melts the desired portion of the powder. The build platform is lowered, and a new layer of powder is deposited using the powder delivery system, as illustrated in Fig. 1.

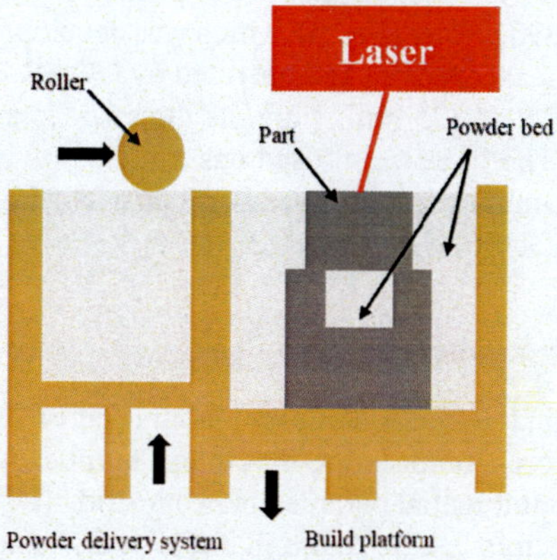

Fig. 1. SLM process

In the process, several thermal phenomena come into play, including melting, conduction, convection, radiation, and evaporation. These phenomena can be observed and understood through Fig. 2, which provides an illustration of their interactions.

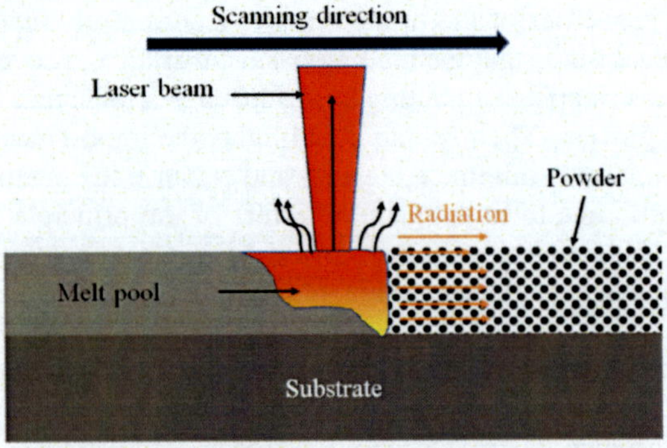

Fig. 2. Phenomena involved

3 Experimental Study

The design of experiments plays a crucial role in organizing scientific research tests to gather a maximum amount of information with a minimum number of experiments regarding the impact of parameter variations on one or more responses. The objective of the design of experiments methodology is to establish relationships between causes and effects, enabling the optimization of the process parameters under investigation. Indeed, this experimental study aims to develop a design of experiments to investigate the influence of the main printing parameters on the on the roughness of the top surface and the construction surface of 316L stainless steel printed parts. The parts were manufactured using the EOS M 290 printer (Fig. 3), which was developed by the German manufacturer EOS (Electro Optical Systems) and operated by UQAR. The printer has a total printing volume of 250 * 250 * 325 mm, while its dimensions are 250 * 250 * 325 mm and it weighs 1250 kg. The laser beam used has a radius of 100 µm, a maximum power of 400 W, and a maximum scanning speed of 7 m/s. The Fig. 4 shows the bilinear strategy adopted for manufacturing these parts.

3.1 Design of Experiments

In our case, a complete factorial design composed of three factors at three levels was chosen in order to encompass the maximum number of combinations. These factors represent the most influential parameters in our study (Di Wang et al. 2016). The specific levels for these factors can be found in Table 1. By generating a three-factorial design with three levels, we obtain a total of 27 combinations, which are listed in Table 2. It's worth noting that the layer thickness is fixed at 20 µm.

3.2 Part Printing

For this study, a prismatic specimens were made with dimensions of 10 * 10 * 14 mm. Four specimens were produced for each configuration, resulting in a total of 108 specimens. This approach was taken to ensure the repeatability of the measurements.

Fig. 3. Selective laser melting 3D printer EOS M290

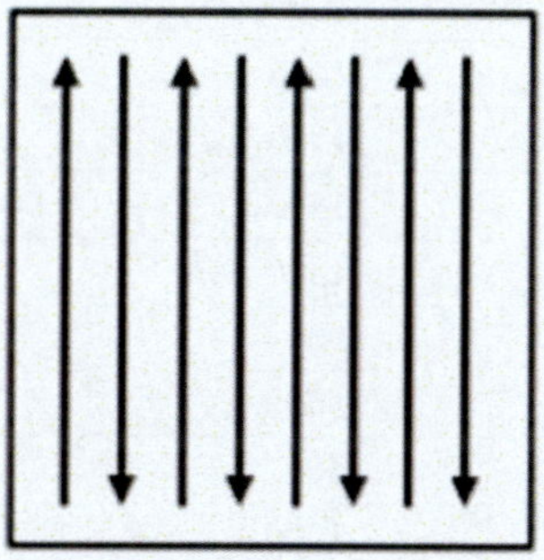

Fig. 4. Bilinear strategy

Table 1. Factors levels

Factor	Level		
Power P(W)	100	200	300
Scanning speed V(mm/s)	500	1000	1500
Hatching distance HD (μm)	100	150	200

During the printing process (as shown in Fig. 5), there was a friction of the scraper blade with parts 10, 19, 20, 21, and 22 due to excessive fusion. As a result, the production of these parts was halted after a few layers of printing. The manufacturing of specimens lasted a total of 25 h. The Fig. 6 represents the printed parts.

Table 2. Design of experiments

Configuration	Power P (W)	scanning speed V (mm/s)	Hatching distance HD(μm)
1	100	500	100
2	100	500	150
3	100	500	200
4	100	1000	100
5	100	1000	150
6	100	1000	200
7	100	1500	100
8	100	1500	150
9	100	1500	200
10	200	500	100
11	200	500	150
12	200	500	200
13	200	1000	100
14	200	1000	150
15	200	1000	200
16	200	1500	100
17	200	1500	150
18	200	1500	200
19	300	500	100
20	300	500	150
21	300	500	200
22	300	1000	100
23	300	1000	150
24	300	1000	200
25	300	1500	100
26	300	1500	150
27	300	1500	200

3.3 Roughness Measurement

The limited surface quality of parts produced through the SLM process is considered one of its main drawbacks, which restricts its potential applications (Ahmed et al. 2022, Greco et al. 2020).

After manufacturing the SLM parts, the measurements of the roughness were conducted on the top surfaces as well as the surfaces parallel to the direction of construction using a contact roughness meter (Fig. 7).

Fig. 5. Printing parts

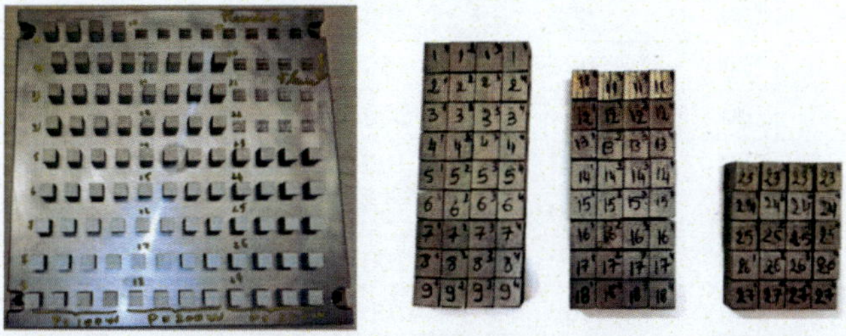

Fig. 6. Printed parts

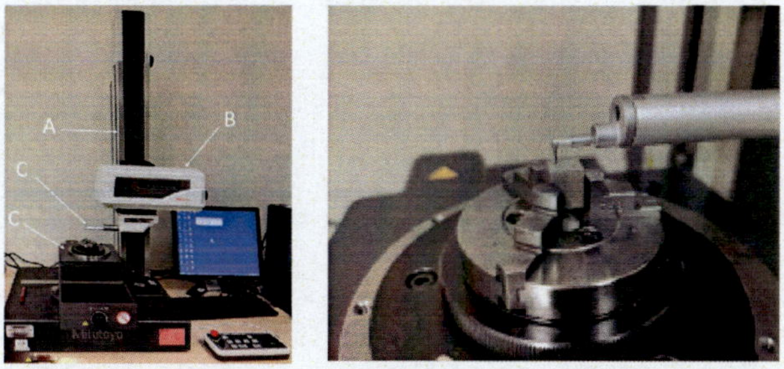

Fig. 7. Contact roughness meter

According to a study by Di Wang et al. (2016), the roughness (Ra) of SLM fabricated parts typically falls within the range of 10 μm to 30 μm. Table 3 presents the roughness measurements conducted on both the top surfaces and built surfaces of each specimen which are well within the margins found in the literature.

Table 3. Roughness measurements'

Conf	P(W)	V(mm/s)	HD(um)	Ra $_{top}$(μm)	Ra $_{built}$(μm)
1	100	500	100	9,22	11,79
2	100	500	150	8,76	10,87
3	100	500	200	9,35	9,72
4	100	1000	100	8,04	10,59
5	100	1000	150	7,17	10,34
6	100	1000	200	9,21	9,51
7	100	1500	100	8,19	11,78
8	100	1500	150	10,92	11,01
9	100	1500	200	12,01	12,17
11	200	500	150	10,83	14,29
12	200	500	200	8,68	11,47
13	200	1000	100	8,54	13,25
14	200	1000	150	9,57	13,28
15	200	1000	200	11,25	13,76
16	200	1500	100	9,97	12,27
17	200	1500	150	8,28	12,05
18	200	1500	200	9,81	12,50
23	300	1000	150	14,81	12,94
24	300	1000	200	13,88	13,96
25	300	1500	100	13,58	13,64
26	300	1500	150	10,26	11,63
27	300	1500	200	10,68	14,56

4 Experimental Results

4.1 Top Surface Roughness

For the top surface roughness, the regression equation modeling Ra_{top} as a function of the parameters P, V and HD is as follows:

$$Ra_{top} = 2.36 - 0.0041\,P + 0.0047\,V + 0.037\,HD + 0.0002\,P^2 + 0.000031\,P * V - 0.000186\,P * HD \quad (1)$$

The results for Ra_{top} are obtained with a coefficient $R^2 = 70.89\%$.

The laser power (P) is identified as the most influential factor on the roughness of the upper surface. Specifically, for a power lower than 220 W, the roughness ranges between 7 and 10 μm. However, for a power higher than 220 W, the roughness increases and ranges between 10 and 14 μm (Fig. 8).

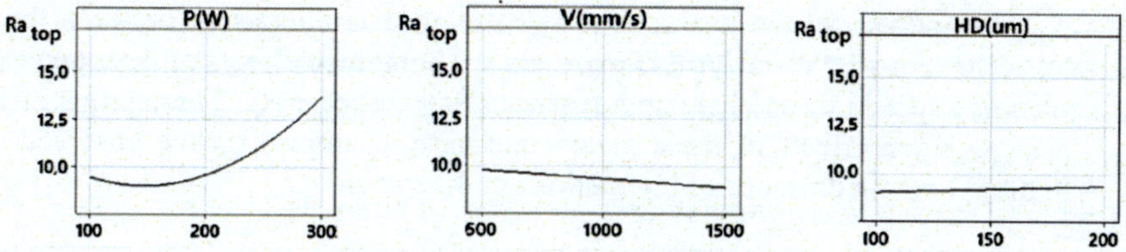

Fig. 8. Main effects graphics for Ra_{top}

The analysis of the interaction diagrams reveals interaction between the parameters. The terms P*V and P*HD represent the combined effects of laser power (P) with scanning speed (V) and hatching distance (HD) respectively. According to Fig. 9a, for $P < 200$ W, $HD = 100\ \mu m$ should be selected to achieve a roughness below 10 μm. As for $P > 200$ W, as HD increases, Ra_{top} decreases. As shown in Fig. 9b, for laser power (P) values below 150 W, it is recommended to use a low scanning speed (V) of 500 mm/s to achieve the best surface quality. However, for laser power values above 150 W, it is advised to increase the scanning speed to 1500 mm/s in order to obtain the optimal surface quality.

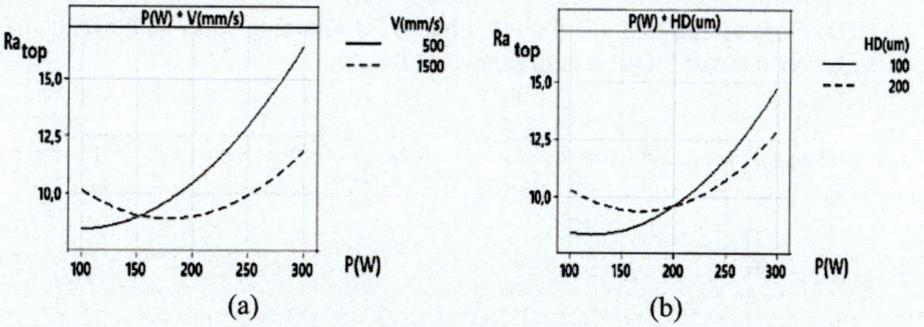

Fig. 9. Interaction diagram for Ra_{top}

4.2 Build Surface Roughness

For the roughness of the building surface, the regression equation modeling the variation of Ra_{built} as a function of P, V and HD is the following:

$$Rabuilt = 12.36 + 0.0335\,P - 0.0019\,V - 0.0365\,HD - 0.000016\,P * V + 0.000029\,V * HD \tag{2}$$

The results for Ra_{built} are obtained with an R^2 coefficient of 71.02%.

The laser power is the primary factor that significantly influences the surface roughness (Ra_{built}) of fabricated objects. Specifically, an increase in power (P) leads to a notable rise in Ra_{built}. In Fig. 10, it is demonstrated that Ra_{built} ranges from 11 μm for a power of 100 W to 14 μm for a power of 300 W.

The analysis of the interaction diagrams reveals interaction between the parameters. The terms P*V and V*HD represent the combined effects of laser power (P) with scanning speed (V) and hatching distance (HD) respectively. These terms indicate that the combined effects of these parameters have an impact on the observed outcome, highlighting the presence of interactions in the system.

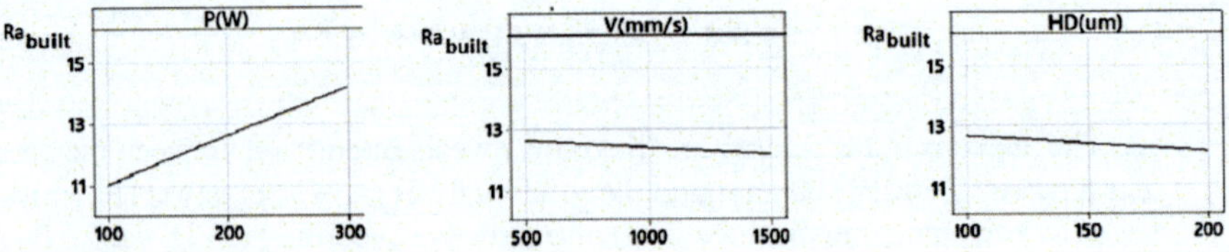

Fig. 10. Main effects graphics for Ra_{built}

In Fig. 11a, for laser power values below 150 W, it is recommended to use a low scanning speed of 500 mm/s. However, for laser power values above 150 W, increasing the scanning speed can help minimize the Ra_{built} parameter. According to Fig. 11b, when the scanning speed is lower than 1250 mm/s, it is advisable to choose a hatching distance (HD) close to 200 μm. On the other hand, for scanning speeds higher than 1250 mm/s, selecting a lower HD can help minimize Ra_{built}.

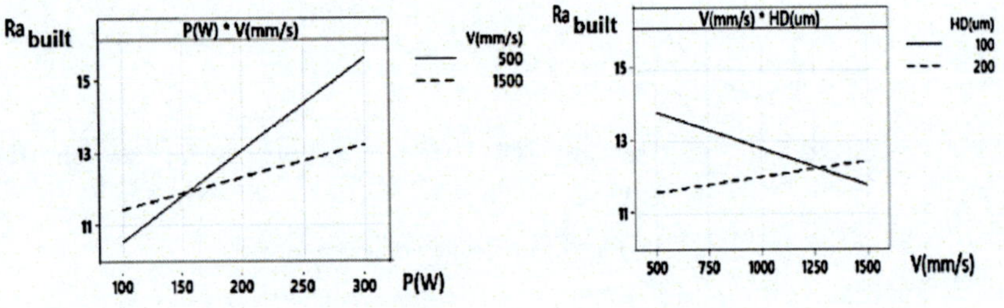

(a) (b)

Fig. 11. Interaction diagram for Ra_{built}

5 Conclusion

This study conducts a parametric analysis to examine the impact of laser power, scanning speed, and hatch distance on the surface roughness of stainless steel parts produced through selective laser melting (SLM). The study findings indicate that, across all configurations, the quality of the top surface consistently surpasses that of the build surfaces.

Laser power (P) emerges as the most influential factor affecting the roughness of both the top and build surfaces. Specifically, increasing the laser power leads to an elevation in roughness, thereby deteriorating the surface condition.

A slight increase in the hatching distance (HD) contributes to a rise in Ra_{top} while reducing Ra_{build}. Furthermore, augmenting the scanning speed aids in enhancing the overall quality of all surfaces.

References

Hao, L., Savalani, M.M., Zhang, Y., Tanner, K.E., Harris, R.A.: Selective laser sintering of hydroxyapatite reinforced polyethylene composites for bioactive implants and tissue scaffold development. Proc. Inst. Mech. Eng. Part H J. Eng. Med. **220**(4), 521–531 (2006). https://doi.org/10.1243/09544119JEIM67

Li, R., Liu, J., Shi, Y., Du, M., Xie, Z.: 316L stainless steel with gradient porosity fabricated by selective laser melting. J. Mater. Eng. Perform. **19**(5), 666–671 (2010). https://doi.org/10.1007/s11665-009-9535-2

Miranda, G., et al.: Predictive models for physical and mechanical properties of 316l stainless steel produced by selective laser melting. Mater. Sci. Eng. A **657**, 43–56 (2016). https://doi.org/10.1016/j.msea.2016.01.028

Di Wang, Y., Yongqiang, Y., Dongming, X.: Theoretical and experimental study on surface roughness of 316l stainless steel metal parts obtained through selective laser melting. Rapid Prototyp. J. **22**(4), 706–716 (2016). https://doi.org/10.1108/RPJ-06-2015-0078

Tucho, W.M., Lysne, V.H., Austbø, H., Sjolyst-Kvernelandn, A., Hansen, V.: Investigation of effects of process parameters on microstructure and hardness of SLM manufactured SS316L. J. Alloy. Compd. **740**, 910–925 (2018). https://doi.org/10.1016/j.jallcom.2018.01.098

Greco, S., Gutzeit, K., Hotz, H., Kirsch, B., Aurich, J.C.: Selective laser melting (SLM) of AISI 316L—impact of laser power, layer thickness, and hatch spacing on roughness, density, and microhardness at constant input energy density. Int. J. Adv. Manuf. Technol. **108**, 1551–1562 (2020). https://doi.org/10.1007/s00170-020-05510-8

Ahmed, N., Barsoum, I., Haidemenopoulos, G., Al-Rub, R.A.: Process parameter selection and optimization of laser powder bed fusion for 316L stainless steel: a review. J. Manuf. Process. **75**, 415–434 (2022). https://doi.org/10.1016/j.jmapro.2021.12.064

Experimental Study of Morphological Defects Generated by SLM on 17-4PH Stainless Steel

Thabet A. M. Sghaier[1] (✉), Habib Sahlaoui[1], Haifa Sallem[2], Tarek Mabrouki[3], and Joël Rech[4]

[1] National High School of Engineering of Tunis (ENSIT), Laboratory of Mechanics, Materials, and Processes (LMMP) LR 99ES05, University of Tunis (UT), 5 Avenue Taha Hussien, B.P 56, 1008 Bab Menara, Tunisia
thabetsgaier@gmail.com, habib.sahlaoui@ensit.rnu.tn

[2] Powder Technology and Advanced Materials, HEI-VS, HES-SO University of Applied Sciences and Arts Western Switzerland, Rue de l'Industrie 23, 1950 Sion, Switzerland
haifa.sallem@hevs.ch

[3] National Engineering School of Tunis (ENIT), Applied Mechanics and Engineering Laboratory (LMAI-LR-11ES19), University of Tunis El Manar, BP-37, Le Belvedere, 1002 Tunis, Tunisia
tarek.mabrouki@enit.utm.tn

[4] Laboratoire de Tribologie et Dynamique Des Système (LTDS-UMR5513), ECL-ENISE, 58, Rue de Jean Parot, 42023 Saint-Etienne Cedex 2, France
joel.rech@enise.fr

Abstract. For cyclic loading applications and critical structural, the emergence of flaws in additively manufactured components is a key concern. This study highlighted the defects generated by selective laser melting (SLM) to the 17-4 PH stainless steel parts. The defects were quantified using an experimental study: Optical microscope and roughness meter to evaluate surface morphology (roughness and porosity) and microstructure. Tensile tests to evaluate the mechanical performance. The defects recorded take several forms, the most important of which are the rough and heterogeneous surface, prominent bumps on the outer surface in the shape of a hemisphere, cavities inside the spaced and adjacent parts of different shapes, internal cracks perpendicular to the printing direction, undissolved particles confined in the dissolving structure layers, brittle behaviour, non-resistant attractive and heterogeneous structure etc.

Keywords: Selective Laser Melting · 17-4PH stainless steel · morphological defects · roughness

1 Introduction

Additive Manufacturing (AM) is also referred to as layer manufacturing, freeform fabricating, or additive techniques, brings together a family of shaping processes that break with those used conventionally. It allows exceptional creative freedom by generating

T. Mabrouki et al. (Eds.): AIAM 2023, LNME, pp. 238–244, 2024.
https://doi.org/10.1007/978-3-031-47784-3_28

previously unthinkable geometries (hollow structures, metal lattices, etc.), guaranteeing much shorter manufacturing times than what can exist in the field of foundry and reducing the material cost compared to machining (Boschetto et al. 2020).

Another particularity of AM is the fast melting-solidification kinetics involved which leads to the appearance of internal and external defects (Sghaier et al. 2023). These defects can be mechanical (Yadollahi et al. 2016; Ponnusamy et al. 2021), morphological (Chaus et al. 2023) and microstructural (Auguste et al. 2018; Leo et al. 2019), weakening the material's mechanical strength and leading to part fragility.

Porosity is a defect found in all metal additive manufacturing processes. They are microscopic spherical voids, around 100 μm in size, that are closed beneath the surface of the part. Porosity rates vary from 1% to 5% and can cause catastrophic failure (Kim and Moylan 2018). There are two types of pores: irregularly shaped pores and spherical pores. The former is formed by a lack of molten metal filling the space. The latter are formed by gas entrained in the powder. Irregularly shaped pores can be created if the heat input is insufficient, leading to incomplete fusion. Incomplete fusion is caused by poor choice of treatment parameters. The size of the pores also depends on these parameters. Investigated the mechanical and microstructural characteristics of 17-4PH SLMed samples, showing that the printed samples exhibit a quasi-cleavage failure mode consisting of microvoid coalescence and brittle fracture, and that the fracture surfaces contain deep quasi-cleavage dimples, especially around inclusions and unmelted particles.

The study presented by Chaus et al. (Chaus et al. 2023) investigated the influence of surface topography and surface roughness on the microstructure evolution of additively manufactured 17-4PH steels. They confirmed that the type of surface topography and therefore the surface roughness of the additively manufactured samples plays an important role in the microstructure evolution during 3D-AM printing.

Thus, after this study, we can say that the mechanical properties are not only influenced by the pores inside the printed material, but also by the morphological nature and roughness of the manufactured material, as they directly influence the microstructural development of the material.

It is in this context that we propose this modest contribution to identify the morphological and microstructural defects generated by SLM in the additively manufactured 17-4 PH stainless steel.

2 Experimental Procedures

In this study, 17-4PH precipitation hardening stainless steel was used to evaluate the defects of 3D printing by SLM. A commercially available 17-4PH powder with a diameter of 15-45 μm was used. The chemical composition of the powder (in wt.%) is given in Table 1 (ASTM A564/A564M).

The selective laser melting of the specimens is carried out on a PHENIX SYSTEMS type PROX200 machine. The manufacturing chamber is maintained under a protected atmosphere (Nitrogen N_2).

All samples were processed with a 10 mm hexagonal lasering strategy in parallel lines forming crosshatches between each fused layer. A rotation with an angle of 90° was performed (odd layer 0°, even layer 90°) with an overlap of the hexagons of 100 μm.

Table 1. Chemical composition of the 17-4PH powder (wt%).

Composition (wt%)	C	Si	Mn	Ni	Cr	Cu	Co	Mo	Nb + Ta
17-4PH	<0.07	<1	<1	3–5	15–17.5	3–5	0.03	<0.5	0.15–0.45

Figure 1 shows the path followed by the laser beam during SLM manufacturing for the used laser scanning strategy.

Table 2. SLM processing parameters

$E = 64\ Jmm^{-3}$	
SLM parameters	Value
Power (W)	240
Laser Scan speed (mm/s)	2500
Hatching distance (μm)	50
Layer thickness (μm)	30
Overlap (%)	30

Two types of specimens (Rectangular plate specimens with dimensions of 12 × 12 × 57 mm^3 and cylindrical with dimensions Ø12 mm × 65 mm) were produced by SLM in the same tray, using a PROX200 machine, with one set of optimized processing parameters highlighted in Table 2.

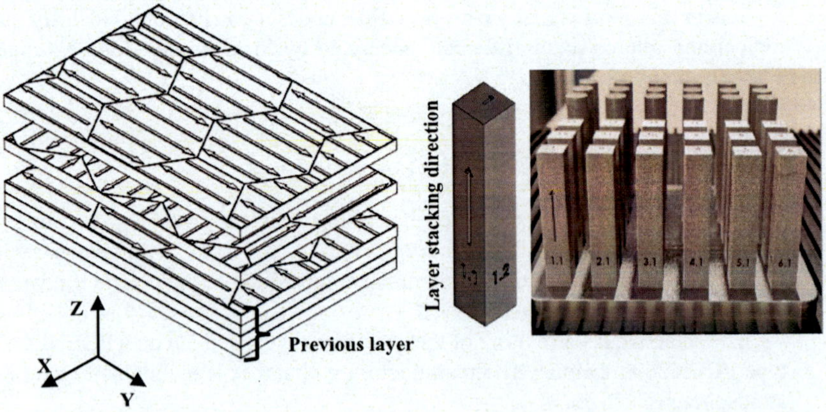

Fig. 1. Laser scanning strategies used in SLM and manufactured test samples.

After production and before disassembly, the parts must be stress-relief annealed at 760° for 2 h, followed by slow air cooling, to minimise residual stresses in the structure and reduce the risk of dimensional change during subsequent machining. Finally, the specimens are cut from the production plate using electro-discharge machining (EDM). All specimens produced by SLM are machined by turning until the standardised shape is achieved for the tensile specimens.

3 Results and Discussion

3.1 Surface Topography and Roughness

The surfaces of the parts produced by SLM are characterised by high roughness parameters (linear: Ra = 13.7 μm and surface: Sa = 16.3 μm) and spherical and domed shapes of various sizes adhering to the lateral surface, as shown in Table 3 and Fig. 2.

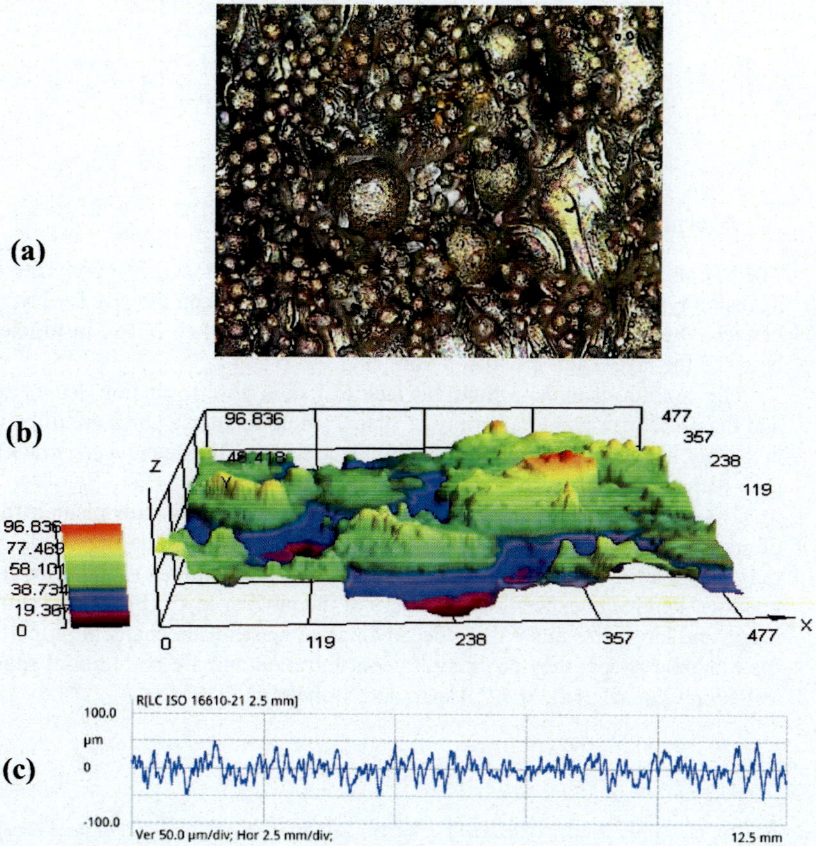

Fig. 2. (a) Surface topography, (b) 3D views of the surface and (c) the results of surface profile measurements extracted from the 17-4PH SLMed samples areas

It is also possible to see powder particles adhering to these shapes, forming satellites, mainly due to the attraction of materials dissolved by the Marangoni effect of the powder surrounding the part.

Dissolved material can also be seen in the form of twisted waves that appear on the outer surface of the piece parallel to the axis of construction, and then hide under bumps and spherical shapes. The lava takes on random distortions that appear as prominent bumps on the surface or a hole as deep as the thickness of the powder layer, and it can also pass through several layers in the same direction of build as shown in Fig. 2a.

Table 3. Linear and surface roughness

Linear (μm)		Surface (μm)	
Ra	13.7	Sa	16.3
Rt	104.9	Sk	47.5
Rz	90.8	Sz	136.8
Rp	43.3	Sp	74.9
Rq	17.3	Sq	21.4

3.2 Morphological Analysis

The morphological analysis of the porosity shows that most of the pores are of the lack of fusion type due to the lack of energy, especially between the powder layers and also between the hatch spacings (Nassar et al. 2014; Sames et al. 2016), insufficient overlap between the layers and the fusion lines (Fig. 3a, b and c).

The conditions under which this lack of fusion pore formation occurs are variable and therefore give rise to a variety of shapes, although these pores are often angular, as is the case when they are associated with delamination between layers, cracks along the melt baths or unmelted powder grains.

The growth of the melt pool during processing can occasionally result in the presence of unmelted powder particles within the lack of fusion area. The size of these pores can range from a few hundred nanometres to several micrometres (100μm) wide (Fig. 3a) and their depth can exceed the thickness of the powder layer (Fig. 3b). Their particular shape and large size make them detrimental, at least for mechanical properties: due to their angular shape, they cause stress concentration and are preferential sites for crack initiation (Kudzal et al. 2017; Asgari and Mohammadi 2018).

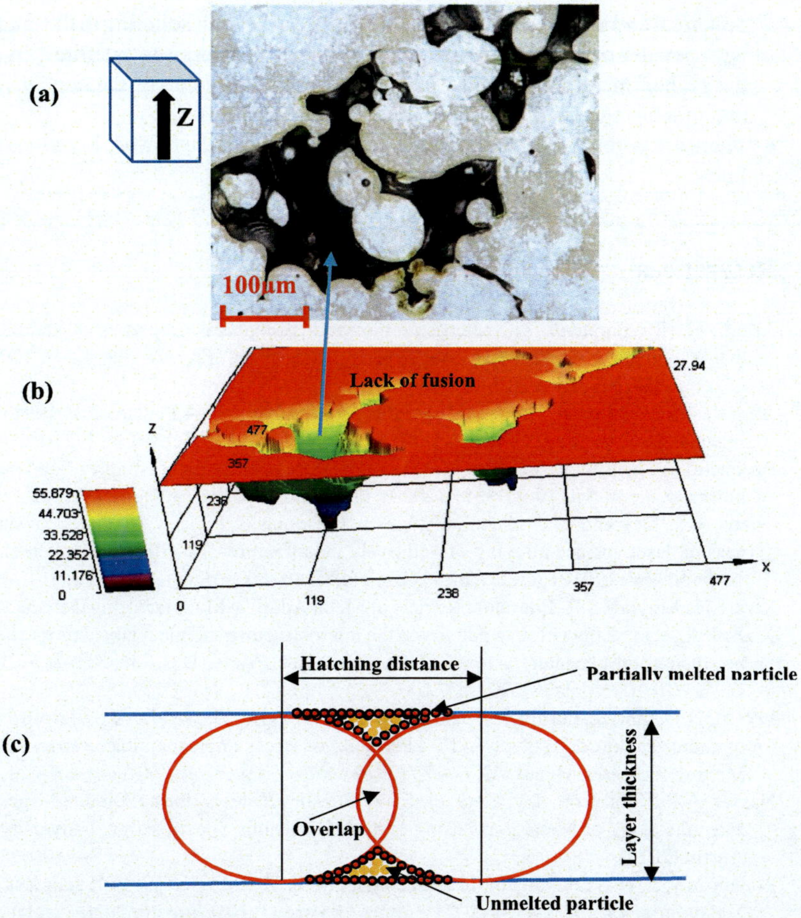

Fig. 3. Porosity morphologies: a) porosity due to lack of fusion, b) 3d image of porosity, and insufficient overlap between the layers and the fusion lines.

4 Conclusion

Identifying defects and understanding their occurrence in the additive manufacturing process is an absolute necessity if we are to work towards improving the quality of 3D printed parts, and for this reason this study focuses on defects arising from the SLM manufacturing process for 17-4PH stainless steel. Therefore, the following conclusions can be drawn from this study:

- The surfaces of the parts produced by SLM are characterised by high linear and surface roughness parameters, especially those that influence fatigue resistance, i.e., Rt=104.9 μm, SP= 136.8 and Sq=21.4 μm.

- Spherical and domed shapes of various sizes can be seen adhering to the lateral surface, with powder particles adhering to these shapes and forming satellites. In addition to the molten material, twisted waves appear on the outside of the part, parallel to the manufacturing axis.
- There are pores caused by the lack of fusion between the powder layers and the fusion lines.

References

Asgari, H., Mohammadi, M.: Microstructure and mechanical properties of stainless steel CX manufactured by direct metal laser sintering. Mater. Sci. Eng. A **709**, 82–89 (2018). https://doi.org/10.1016/j.msea.2017.10.045

Auguste, P., Mauduit, A., Fouquet, L., Pillot, S.: Study on 17-4 PH stainless steel produced by selective laser melting. UPB Sci. Bull. Ser. B Chem. Mater. Sci. **80**, 197–210 (2018)

Boschetto, A., Bottini, L., Macera, L., Veniali, F.: Post-processing of complex SLM parts by barrel finishing. Appl. Sci. **10**(4), 1382 (2020). https://doi.org/10.3390/app10041382

Chaus, A.S., Devoino, O.G., Sahul, M., Vančo, Ľ, Buranský, I., Kusý, M.: Effect of surface roughness on laser surface alloying of additively manufactured 17-4PH stainless steel. Surf. Coat. Technol. **454**, 129161 (2023). https://doi.org/10.1016/j.surfcoat.2022.129161

Kim, F.H., Moylan, S.P.: Literature Review of Metal Additive Manufacturing Defects. NIST (2018)

Kudzal, A., et al.: Effect of scan pattern on the microstructure and mechanical properties of powder bed fusion additive manufactured 17-4 stainless steel. Mater. Des. **133**, 205–215 (2017). https://doi.org/10.1016/j.matdes.2017.07.047

Leo, P., D'Ostuni, S., Perulli, P., Castro Sastre, M.A., Fernández-Abia, A.I., Barreiro, J.: Analysis of microstructure and defects in 17-4 PH stainless steel sample manufactured by Selective Laser Melting. Procedia Manuf. **41**, 66–73 (2019). https://doi.org/10.1016/j.promfg.2019.07.030

Nassar, A.R., Spurgeon, T.J., Reutzel, E.W.: Sensing Defects during Directed-Energy Additive Manufacturing of Metal Parts using Optical Emissions Spectroscopy. University of Texas at Austin (2014)

Ponnusamy, P., et al.: A study of tensile behavior of SLM processed 17-4 PH stainless steel. Mater. Today Proc. **45**, 4531–4534 (2021). https://doi.org/10.1016/j.matpr.2020.12.1104

Sames, W.J., List, F.A., Pannala, S., Dehoff, R.R., Babu, S.S.: The metallurgy and processing science of metal additive manufacturing. Int. Mater. Rev. **61**(5), 315–360 (2016). https://doi.org/10.1080/09506608.2015.1116649

Sghaier, T.A.M., Sahlaoui, H., Mabrouki, T., et al.: Selective laser melting of stainless-steel: a review of process, microstructure, mechanical properties and post-processing treatments. Int. J. Mater. Form. **16**, 41 (2023). https://doi.org/10.1007/s12289-023-01769-w

Yadollahi, A., Shamsaei, N., Thompson, S.M., Elwany, A., Bian, L.: Mechanical and Microstructural Properties of Selective Laser Melted 17-4 PH Stainless Steel. American Society of Mechanical Engineers Digital Collection (2016)

Author Index

T. Mabrouki et al. (Eds.): AIAM 2023, LNME, pp. 245–246, 2024.
https://doi.org/10.1007/978-3-031-47784-3